Two Manuals Combined into One Book

Electronic Warfare in Operations

And

Military Role in Civil Disturbance Operations

Department of Defense

Pentagon
Publishing

Field Manual
No. 3-36

Headquarters
Department of the Army
Washington, DC, 25 February 2009

Electronic Warfare in Operations

Contents

Figures

Preface

PURPOSE

FM 3-36 provides Army doctrine for electronic warfare (EW) planning, preparation, execution, and assessment in support of full spectrum operations. Users of FM 3-36 must be familiar with full spectrum operations established in FM 3-0; the military decisionmaking process established in FM 5-0; the operations process established in FMI 5-0.1; commander's visualization described in FM 6-0; and electronic warfare described in JP 3-13.1.

SCOPE

FM 3-36 is organized into seven chapters and six appendixes. Each chapter addresses a major aspect of Army EW operations. The appendixes address aspects of EW operations that complement the operational doctrine. A glossary contains selected terms.

- Chapter 1 discusses the nature and scope of electronic warfare and the impact of the electromagnetic environment on Army operations.
- Chapter 2 offers a discussion of EW support to full spectrum operations, combat power, the warfighting functions, and information tasks.
- Chapter 3 introduces the organizational framework for command and control of EW operations.
- Chapter 4 describes how commanders integrate EW operations throughout the operations process.
- Chapter 5 discusses the coordination required to synchronize and deconflict EW operations effectively.
- Chapter 6 provides the baseline for integrating EW operations into joint and multinational operations.
- Chapter 7 discusses the enabling activities that support EW operations, such as command and control, intelligence, logistics, technical support and EW training.
- Appendix A discusses the electromagnetic environment.
- Appendix B illustrates an EW appendix to an operation order.
- Appendix C illustrates an EW running estimate.
- Appendix D discusses EW related reports and messages.
- Appendix E offers a reference guide to Army and joint EW capabilities.
- Appendix F discusses EW-related tools and resources.

APPLICABILITY

FM 3-36 provides guidance on EW operations for commanders and staffs at all echelons. This FM serves as an authoritative reference for personnel who—

- Develop doctrine (fundamental principles and tactics, techniques, and procedures), materiel, and force structure.
- Develop institutional and unit training.
- Develop standing operating procedures for unit operations.
- Conduct planning, preparation, execution and assessment of electronic warfare.

FM 3-36 applies to the Active Army, Army National Guard/Army National Guard of the United States, and U.S. Army Reserve, unless otherwise stated.

Chapter 1

Electronic Warfare Overview

This chapter provides an overview of electronic warfare and the conceptual foundation that leaders require to understand the electromagnetic environment and its impact on Army operations.

OPERATIONAL ENVIRONMENTS

1-1. An *operational environment* is a composite of the conditions, circumstances, and influences that affect the employment of capabilities and bear on the decisions of the commander (JP 3-0). An operational environment includes physical areas—the air, land, maritime, and space domains. It also includes the information that shapes the operational environment as well as enemy, adversary, friendly, and neutral systems relevant to a joint operation. Joint planners analyze operational environments in terms of six interrelated operational variables: political, military, economic, social, information, and infrastructure. To these variables Army doctrine adds two more: physical environment and time. (See FM 3-0 for additional information on the operational variables). Army leaders use operational variables to understand and analyze the broad environment in which they are conducting operations.

1-2. Army leaders use mission variables to synthesize operational variables and tactical-level information with local knowledge about conditions relevant to their mission. They use mission variables to focus analysis on specific elements that directly affect their mission. Upon receipt of a warning order or mission, Army tactical leaders narrow their focus to six mission variables known as METT-TC. They are mission, enemy, terrain and weather, troops and support available, time available and civil considerations. The mission variables outline the situation as it applies to a specific Army unit.

1-3. Commanders employ and integrate their unit's capabilities and actions within their operational environment to achieve a desired end state. Through analyzing their operational environment, commanders understand how the results of friendly, adversary, and neutral actions may impact that end state. During military operations, both friendly and enemy commanders depend on the flow of information to make informed decisions. This flow of information depends on the electronic systems and devices used to communicate, navigate, sense, store, and process information.

INFORMATION AND THE ELECTROMAGNETIC SPECTRUM

1-4. Commanders plan for and operate electronic systems and the weapon systems that depend on them in an intensive and nonpermissive electromagnetic environment. They ensure the flow of information required for their decisionmaking. (Appendix A further discusses the electromagnetic environment.) Within the electromagnetic environment, electronic systems and devices operate in the electromagnetic spectrum. (See figure 1-1, page 1-2.)

1-5. The electromagnetic spectrum has been used for commercial and military applications for over a century. However, the full potential for its use as the primary enabler of military operations is not yet fully appreciated. New technologies are expanding beyond the traditional radio frequency spectrum. They include high-power microwaves and directed-energy weapons. These new technologies are part of an electronic warfare (EW) revolution by military forces. Just as friendly forces leverage the electromagnetic spectrum to their advantage, so do capable enemies use the electromagnetic spectrum to threaten friendly force operations. The threat is compounded by the growth of a wireless world and the increasingly sophisticated use of commercial off-the-shelf technologies.

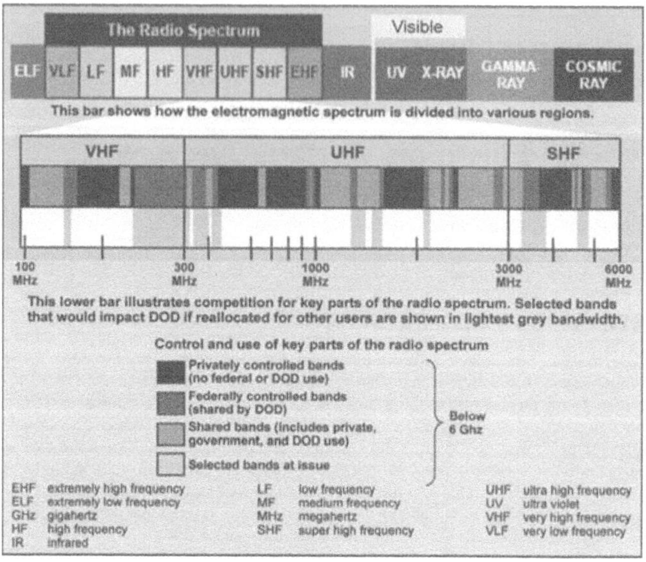

Figure 1-1. The electromagnetic spectrum

1-6. Adversaries and enemies, from small and single actors to large state, multinational, and nonstate actors, use the most modern technology. Such technology is moving into the cellular and satellite communications area. Most military and commercial operations rely on electromagnetic technologies and are susceptible to the inherent vulnerabilities associated with their use. This reliance requires Army forces to dominate the electromagnetic spectrum (within their operational environment) with the same authority that they dominate traditional land warfare operations. Emerging electromagnetic technologies offer expanded EW capabilities. They dynamically affect the electromagnetic spectrum through delivery and integration with other types of emerging weapons and capabilities. Examples are directed-energy weapons, high-powered microwaves, lasers, infrared, and electro-optical and wireless networks and devices.

1-7. In any conflict, commanders attempt to dominate the electromagnetic spectrum. They do this by locating, targeting, exploiting, disrupting, degrading, deceiving, denying, or destroying the enemy's electronic systems that support military operations or deny the spectrum's use by friendly forces. The increasing portability and affordability of sophisticated electronic equipment guarantees that the electromagnetic environment in which forces operate will become even more complex. To ensure unimpeded access to and use of the electromagnetic spectrum, commanders plan, prepare, execute, and assess EW operations against a broad set of targets within the electromagnetic spectrum. (See figure 1-2.)

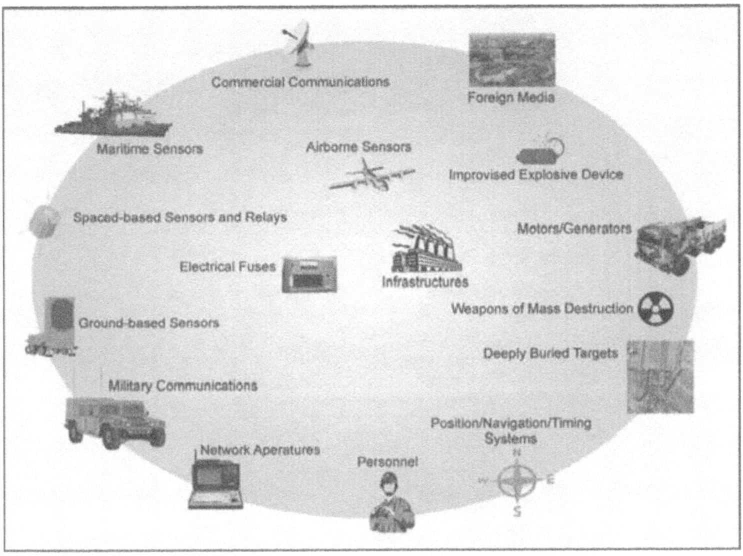

Figure 1-2. Electromagnetic spectrum targets

DIVISIONS OF ELECTRONIC WARFARE

1-8. *Electronic warfare* is defined as military action involving the use of electromagnetic and directed energy to control the electromagnetic spectrum or to attack the enemy. Electronic warfare consists of three divisions: electronic attack, electronic protection, and electronic warfare support (JP 3-13.1). (See figure 1-3.)

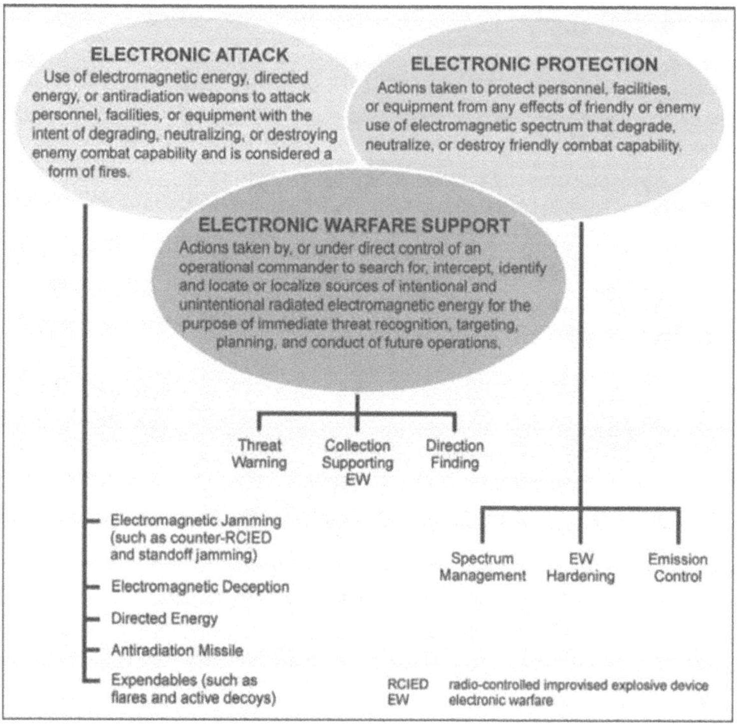

Figure 1-3. The three subdivisions of electronic warfare

ELECTRONIC ATTACK

1-9. *Electronic attack* is a division of electronic warfare involving the use of electromagnetic energy, directed energy, or antiradiation weapons to attack personnel, facilities, or equipment with the intent of degrading, neutralizing, or destroying enemy combat capability and is considered a form of fires (JP 3-13.1). Electronic attack includes—

- Actions taken to prevent or reduce an enemy's effective use of the electromagnetic spectrum, such as jamming and electromagnetic deception.
- Employment of weapons that use either electromagnetic or directed energy as their primary destructive mechanism (lasers, radio frequency weapons, particle beams).
- Offensive and defensive activities including countermeasures.

1-10. Common types of electronic attack include spot, barrage, and sweep electromagnetic jamming. Electronic attack actions also include various electromagnetic deception techniques such as false target or duplicate target generation. (See paragraphs 1-23 to 1-31 for further discussion of electronic attack activities.)

1-11. *Directed energy* is an umbrella term covering technologies that relate to the production of a beam of concentrated electromagnetic energy or atomic or subatomic particles (JP 1-02). A directed-energy weapon uses directed energy primarily as a direct means to damage or destroy an enemy's equipment, facilities, and personnel. In addition to destructive effects, directed-energy weapon systems support area denial and crowd control. (See appendix A for more information on directed energy.)

1-12. Examples of offensive electronic attack include—

- Jamming enemy radar or electronic command and control systems.
- Using antiradiation missiles to suppress enemy air defenses (antiradiation weapons use radiated energy emitted from the target as their mechanism for guidance onto targeted emitters).
- Using electronic deception techniques to confuse enemy intelligence, surveillance, and reconnaissance systems.
- Using directed-energy weapons to disable an enemy's equipment or capability.

1-13. Defensive electronic attack uses the electromagnetic spectrum to protect personnel, facilities, capabilities, and equipment. Examples include self-protection and other protection measures such as use of expendables (flares and active decoys), jammers, towed decoys, directed-energy infrared countermeasure systems, and counter-radio-controlled improvised-explosive-device systems. (See JP 3-13.1 for more discussion of electronic attack.)

ELECTRONIC PROTECTION

1-14. *Electronic protection* is a division of electronic warfare involving actions taken to protect personnel, facilities, and equipment from any effects of friendly or enemy use of the electromagnetic spectrum that degrade, neutralize, or destroy friendly combat capability (JP 3-13.1). For example, electronic protection includes actions taken to ensure friendly use of the electromagnetic spectrum, such as frequency agility in a radio, or variable pulse repetition frequency in radar. Electronic protection should not be confused with self-protection. Both defensive electronic attack and electronic protection protect personnel, facilities, capabilities, and equipment. However, electronic protection protects from the effects of electronic attack (friendly and enemy), while defensive electronic attack primarily protects against lethal attacks by denying enemy use of the electromagnetic spectrum to guide or trigger weapons.

1-15. During operations, electronic protection includes, but is not limited to, the application of training and procedures for countering enemy electronic attack. Army commanders and forces understand the threat and vulnerability of friendly electronic equipment to enemy electronic attack and take appropriate actions to safeguard friendly combat capability from exploitation and attack. Electronic protection measures minimize the enemy's ability to conduct electronic warfare support (electronic warfare support is discussed in paragraphs 1-18 to 1-20) and electronic attack operations successfully against friendly forces. To protect friendly combat capabilities, units—

- Regularly brief force personnel on the EW threat.
- Ensure that electronic system capabilities are safeguarded during exercises, workups, and predeployment training.
- Coordinate and deconflict electromagnetic spectrum usage.
- Provide training during routine home station planning and training activities on appropriate electronic protection active and passive measures.
- Take appropriate actions to minimize the vulnerability of friendly receivers to enemy jamming (such as reduced power, brevity of transmissions, and directional antennas).

1-16. Electronic protection also includes spectrum management. The spectrum manager works for the G-6 or S-6 and plays a key role in the coordination and deconfliction of spectrum resources allocated to the force. Spectrum managers or their direct representatives participate in the planning for EW operations.

1-17. The development and acquisition of communications and electronic systems includes electronic protection requirements to clarify performance parameters. Army forces design their equipment to limit inherent vulnerabilities. If electronic attack vulnerabilities are detected, then units must review these programs. (See DODI 4650.01 for information on the spectrum certification process and electromagnetic compatibility.)

Electronic Warfare Support

1-18. *Electronic warfare support* is a division of electronic warfare involving actions tasked by, or under the direct control of, an operational commander to search for, intercept, identify, and locate or localize sources of intentional and unintentional radiated electromagnetic energy for the purpose of immediate threat recognition, targeting, planning, and conduct of future operations (JP 3-13.1).

1-19. Electronic warfare support systems are a source of information for immediate decisions involving electronic attack, electronic protection, avoidance, targeting, and other tactical employments of forces. Electronic warfare support systems collect data and produce information or intelligence to—

- Corroborate other sources of information or intelligence.
- Conduct or direct electronic attack operations.
- Initiate self-protection measures.
- Task weapon systems.
- Support electronic protection efforts.
- Create or update EW databases.
- Support information tasks.

1-20. Electronic warfare support and signals intelligence missions use the same resources. The two differ in the detected information's intended use, the degree of analytical effort expended, the detail of information provided, and the time lines required. Like tactical signals intelligence, electronic warfare support missions respond to the immediate requirements of a tactical commander. Signals intelligence above the tactical level is under the operational control of the National Security Agency and directly supports the overarching national security mission. Resources that collect tactical-level electronic warfare support data can simultaneously collect national-level signals intelligence. See FM 2-0 for more information on signals intelligence.

ACTIVITIES AND TERMINOLOGY

1-21. Although new equipment and tactics, techniques, and procedures continue to be developed, the physics of electromagnetic energy remains constant. Hence, effective EW activities remain the same despite changes in hardware and tactics. Principal EW activities are discussed in the following paragraphs.

PRINCIPAL ACTIVITIES

1-22. Principal EW activities support full spectrum operations by exploiting the opportunities and vulnerabilities inherent in the use of the electromagnetic spectrum. The numerous EW activities are categorized by the EW subdivisions with which they are most closely associated: electronic attack, electronic warfare support, and electronic protection. JP 3-13.1 discusses these principal activities in detail.

Electronic Attack Activities

1-23. Activities related to electronic attack are either offensive or defensive and include—

- Countermeasures.
- Electromagnetic deception.
- Electromagnetic intrusion.
- Electromagnetic jamming.
- Electromagnetic pulse.
- Electronic probing.

Countermeasures

1-24. *Countermeasures* are that form of military science that, by the employment of devices and/or techniques, has as its objective the impairment of the operational effectiveness of enemy activity (JP 1-02). They can be deployed preemptively or reactively. Devices and techniques used for EW countermeasures include electro-optical-infrared countermeasures and radio frequency countermeasures.

1-25. *Electro-optical-infrared countermeasures* consist of any device or technique employing electro-optical-infrared materials or technology that is intended to impair or counter the effectiveness of enemy activity, particularly with respect to precision guided weapons and sensor systems. Electro-optical-infrared is the part of the electromagnetic spectrum between the high end of the far infrared and the low end of ultraviolet. Electro-optical-infrared countermeasures may use laser and broadband jammers, smokes/aerosols, signature suppressants, decoys, pyrotechnics/pyrophorics, high-energy lasers, or directed infrared energy countermeasures (JP 3-13.1).

1-26. *Radio frequency countermeasures* consist of any device or technique employing radio frequency materials or technology that is intended to impair the effectiveness of or counter enemy activity, particularly with respect to precision guided weapons and sensor systems (JP 3-13.1).

Electromagnetic Deception

1-27. *Electromagnetic deception* is the deliberate radiation, reradiation, alteration, suppression, absorption, denial, enhancement, or reflection of electromagnetic energy in a manner intended to convey misleading information to an enemy or to enemy electromagnetic-dependent weapons, thereby degrading or neutralizing the enemy's combat capability (JP 3-13.4). Among the types of electromagnetic deception are the following:

- Manipulative electromagnetic deception involves actions to eliminate revealing, or convey misleading, electromagnetic telltale indicators that may be used by hostile forces.
- Simulative electromagnetic deception involves actions to simulate friendly, notional, or actual capabilities to mislead hostile forces.
- Imitative electromagnetic deception introduces electromagnetic energy into enemy systems that imitates enemy emissions.

Electromagnetic Intrusion

1-28. *Electromagnetic intrusion* is the intentional insertion of electromagnetic energy into transmission paths in any manner, with the objective of deceiving operators or of causing confusion (JP 1-02).

Electromagnetic Jamming

1-29. *Electromagnetic jamming* is the deliberate radiation, re-radiation, or reflection of electromagnetic energy for the purpose of preventing or reducing an enemy's effective use of the electromagnetic spectrum, with the intent of degrading or neutralizing the enemy's combat capability (JP 1-02).

Electromagnetic Pulse

1-30. *Electromagnetic pulse* is the electromagnetic radiation from a strong electronic pulse, most commonly caused by a nuclear explosion that may couple with electrical or electronic systems to produce damaging current and voltage surges (JP 1-02).

Electronic Probing

1-31. *Electronic probing* is the intentional radiation designed to be introduced into the devices or systems of potential enemies for the purpose of learning the functions and operational capabilities of the devices (JP 1-02). This activity is coordinated through joint or interagency channels and supported by Army forces.

Electronic Warfare Support Activities

1-32. Activities related to electronic warfare support include—

- Electronic reconnaissance.
- Electronic intelligence.
- Electronics security.

Electronic Reconnaisance

1-33. *Electronic reconnaissance* is the detection, location, identification, and evaluation of foreign electromagnetic radiations (JP 1-02).

Electronic Intelligence

1-34. *Electronic intelligence* is technical and geolocation intelligence derived from foreign noncommunications electromagnetic radiations emanating from other than nuclear detonations or radioactive sources (JP 1-02).

Electronics Security

1-35. *Electronics security* is the protection resulting from all measures designed to deny unauthorized persons information of value that might be derived from their interception and study of noncommunications electromagnetic radiations, e.g., radar (JP 1-02).

Electronic Protection Activities

1-36. Activities related to electronic protection include—

- Electromagnetic hardening.
- Electromagnetic interference.
- Electronic masking.
- Electronic warfare reprogramming.
- Emission control.
- Spectrum management.
- Wartime reserve modes.
- Electromagnetic compatibility.

Electromagnetic Hardening

1-37. *Electromagnetic hardening* consists of action taken to protect personnel, facilities, and/or equipment by filtering, attenuating, grounding, bonding, and/or shielding against undesirable effects of electromagnetic energy (JP 1-02).

Electromagnetic Interference

1-38. *Electromagnetic interference* is any electromagnetic disturbance that interrupts, obstructs, or otherwise degrades or limits the effective performance of electronics and electrical equipment. It can be induced intentionally, as in some forms of electronic warfare, or unintentionally, as a result of spurious emissions and responses, intermodulation products and the like (JP 1-02).

Electronic Masking

1-39. *Electronic masking* is the controlled radiation of electromagnetic energy on friendly frequencies in a manner to protect the emissions of friendly communications and electronic systems against enemy electronic warfare support measures/signals intelligence, without significantly degrading the operation of friendly systems (JP 1-02).

Electronic Warfare Reprogramming

1-40. *Electronic warfare reprogramming* is the deliberate alteration or modification of electronic warfare or target sensing systems, or the tactics and procedures that employ them, in response to validated changes in equipment, tactics, or the electromagnetic environment. These changes may be the result of deliberate actions on the part of friendly, adversary, or third parties; or may be brought about by electromagnetic interference or other inadvertent phenomena. The purpose of electronic warfare reprogramming is to maintain or enhance the effectiveness of electronic warfare and target sensing system equipment. Electronic warfare reprogramming includes changes to self-defense systems, offensive weapons systems, and intelligence collection systems (JP 3-13.1).

Emission Control

1-41. *Emission control* is the selective and controlled use of electromagnetic, acoustic, or other emitters to optimize command and control capabilities while minimizing transmissions for operations security: a. detection by enemy sensors; b. mutual interference among friendly systems; and/or c. enemy interference with the ability to execute a military deception plan (JP 1-02).

Electromagnetic Spectrum Management

1-42. *Electromagnetic spectrum management* is planning, coordinating, and managing joint use of the electromagnetic spectrum through operational, engineering, and administrative procedures. The objective of spectrum management is to enable electronic systems to perform their functions in the intended environment without causing or suffering unacceptable interference (JP 6-0).

Wartime Reserve Modes

1-43. *Wartime reserve modes* are characteristics and operating procedures of sensors, communications, navigation aids, threat recognition, weapons, and countermeasures systems that will contribute to military effectiveness if unknown to or misunderstood by opposing commanders before they are used, but could be exploited or neutralized if known in advance. Wartime reserve modes are deliberately held in reserve for wartime or emergency use and seldom, if ever, applied or intercepted prior to such use (JP 1-02).

Electromagnetic Compatibility

1-44. *Electromagnetic compatibility* is the ability of systems, equipment, and devices that utilize the electromagnetic spectrum to operate in their intended operational environments without suffering unacceptable degradation or causing unintentional degradation because of electromagnetic radiation or response. It involves the application of sound electromagnetic spectrum management; system, equipment, and device design configuration that ensures interference-free operation; and clear concepts and doctrines that maximize operational effectiveness (JP 1-02).

APPLICATION TERMINOLOGY

1-45. EW capabilities are applied from the air, land, sea, and space by manned, unmanned, attended, or unattended systems. Units employ EW capabilities to achieve the desired lethal or nonlethal effect on a given target. Units maintain freedom of action in the electromagnetic spectrum while controlling the use of it by the enemy. Regardless of the application, units employing EW capabilities must use appropriate levels of control and protection of the electromagnetic spectrum. In this way, they avoid adversely affecting friendly forces. (Improper EW actions must be avoided because they may cause fratricide or eliminate high-value intelligence targets.)

1-46. In the context of EW application, units use several terms to facilitate control and protection of the electromagnetic spectrum. Terms used in EW application include control, detection, denial, deception, disruption and degradation, protection, and destruction. The three subdivisions of EW—electronic attack, electronic protection, and electronic warfare support—are specified within the following descriptions.

Control

1-47. In the context of EW, control of the electromagnetic spectrum is achieved by effectively coordinating friendly systems while countering enemy systems. Electronic attack limits enemy use of the electromagnetic spectrum. Electronic protection secures use of the electromagnetic spectrum for friendly forces, and electronic warfare support enables the commander's accurate assessment of the situation. All three are integrated for effectiveness. Commanders ensure maximum integration of communications; intelligence, surveillance, and reconnaissance; and information tasks.

Detection

1-48. In the context of EW, detection is the active and passive monitoring of the operational environment for radio frequency, electro-optic, laser, infrared, and ultraviolet electromagnetic threats. Detection is the first step in EW for exploitation, targeting, and defensive planning. Friendly forces maintain the capability to detect and characterize interference as hostile jamming or unintentional electromagnetic interference.

Denial

1-49. In the context of EW, denial is controlling the information an enemy receives via the electromagnetic spectrum and preventing the acquisition of accurate information about friendly forces. Degradation uses traditional jamming techniques, expendable countermeasures, destructive measures, or network applications. These range from limited effects up to complete denial of usage.

Deception

1-50. In the context of EW, deception is confusing or misleading an enemy by using some combination of human-produced, mechanical, or electronic means. Through use of the electromagnetic spectrum, EW deception manipulates the enemy's decision loop, making it difficult to establish accurate situational awareness.

Disruption and Degradation

1-51. In the context of EW, disruption and degradation techniques interfere with the enemy's use of the electromagnetic spectrum to limit enemy combat capabilities. This is achieved with electronic jamming, electronic deception, and electronic intrusion. These enhance attacks on hostile forces and act as force multipliers by increasing enemy uncertainty, while reducing uncertainty for friendly forces. Advanced electronic attack techniques offer the opportunity to nondestructively disrupt or degrade enemy infrastructure.

Protection

1-52. In the context of EW, protection is the use of physical properties; operational tactics, techniques, and procedures; and planning and employment processes to ensure friendly use of the electromagnetic spectrum. This includes ensuring that offensive EW activities do not electronically destroy or degrade friendly intelligence sensors or communications systems. Protection is achieved by component hardening, emission control, and frequency management and deconfliction. Frequency management and deconfliction include the capability to detect, characterize, geolocate, and mitigate electromagnetic interference that affects operations. Protection includes other means to counterattack and defeat enemy attempts to control the electromagnetic spectrum. Additionally, organizations such as a joint force commander's EW staff or a joint EW coordination cell enhance electronic protection by deconflicting EW efforts.

Destruction

1-53. Destruction, in the context of EW, is the elimination of targeted enemy systems. Sensors and command and control nodes are lucrative targets because their destruction strongly influences the enemy's perceptions and ability to coordinate actions. Various weapons and techniques ranging from conventional munitions and directed energy weapons to network attacks can destroy enemy systems that use the electromagnetic spectrum. Electronic warfare support provides target location and related information. While destroying enemy equipment can effectively deny the enemy use of the electromagnetic spectrum, the duration of denial will depend on the enemy's ability to reconstitute. (See JP 3-13.1.)

MEANS VERSUS EFFECTS

1-54. EW means are applied against targets to create a full range of lethal and nonlethal effects. (See figure 1-4.) Choosing a specific EW capability depends on the desired effect on the target and other considerations, such as time sensitivity or limiting collateral damage. EW capabilities provide commanders with additional options for achieving their objectives. During major combat operations there may be circumstances where commanders want to limit the physical damage on a given target. Under such circumstances, the EW staff articulates clearly to the commander the lethal and nonlethal effects EW capabilities can achieve. For example, a target might be enemy radar mounted on a fixed tower. Two EW options to defeat the radar could be to jam the radar or destroy it with antiradiation missiles. If the commander desired to limit damage to the tower, an electronic attack jamming platform would be preferred. In circumstances where commanders cannot sufficiently limit undesired effects such as collateral damage, they may be constrained from applying physical force. The EW staff articulates succinctly how EW capabilities can support actions to achieve desired effects and provide lethal and nonlethal options for commanders.

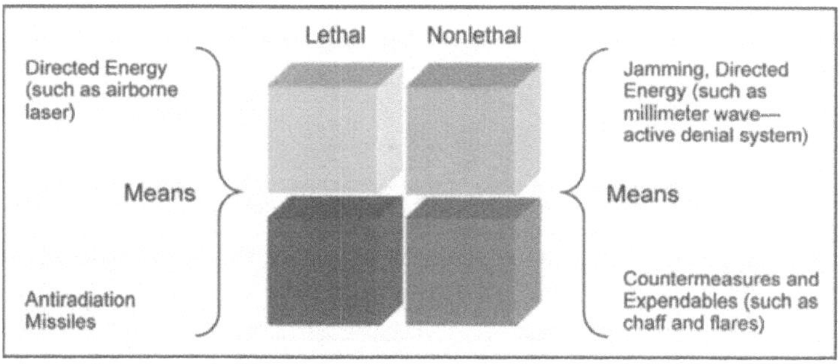

Figure 1-4. Means versus effects

SUMMARY

1-55. As the modern battlefield becomes more technologically sophisticated, military operations continue to be executed in an increasingly complex electromagnetic environment. Therefore, commanders and staffs need to thoroughly understand and articulate how the electromagnetic environment impacts their operations and how friendly EW operations can be used to gain an advantage. Commanders and staffs use the terminology presented in this chapter to describe the application of EW. This ensures a common understanding and consistency within plans, orders, standing operating procedures, and directives.

Chapter 2

Electronic Warfare in Full Spectrum Operations

Information technology is becoming universally available. Most enemies rely on communications and computer networks to make and implement decisions. Radios remain the backbone of tactical military command and control architectures. However, most communications relayed over radio networks are becoming digital as more computers link networks through transmitted frequencies. Therefore, the ability to dominate the electromagnetic spectrum is central to full spectrum operations. This chapter describes how commanders apply electronic warfare capabilities to support full spectrum operations.

THE ROLE OF ELECTRONIC WARFARE

2-1. Army electronic warfare (EW) operations seek to provide the land force commander with capabilities to support full spectrum operations. Full spectrum operations consist of the purposeful, simultaneous combination of offense, defense, and stability or civil support. The goal of full spectrum operations is to change the operational environment so that peaceful processes are dominant. Nonetheless, operational environments are complex; commanders must conduct operations across the entire spectrum of conflict. The Army maintains flexible forces with balanced capabilities and capacities. These flexible and balanced forces remain able to conduct major operations while executing other day-to-day smaller-scale operations. (See FM 3-0.)

2-2. Figure 2-1 (page 2-2) shows the weight of effort for using EW during operations. This figure adapts the elements of full spectrum operations (offense, defense, and stability or civil support) as described in FM 3-0. Overseas, Army forces conduct full spectrum operations (offensive, defensive, and stability) simultaneously as part of a joint force. Within the United States, Army forces conduct homeland defense and civil support operations as part of homeland security. Army electronic warfare (EW) operations seek to provide the land force commander with capabilities to support full spectrum operations. As noted in figure 2-1, statutory law limits the use of EW capabilities in support of civil support operations.

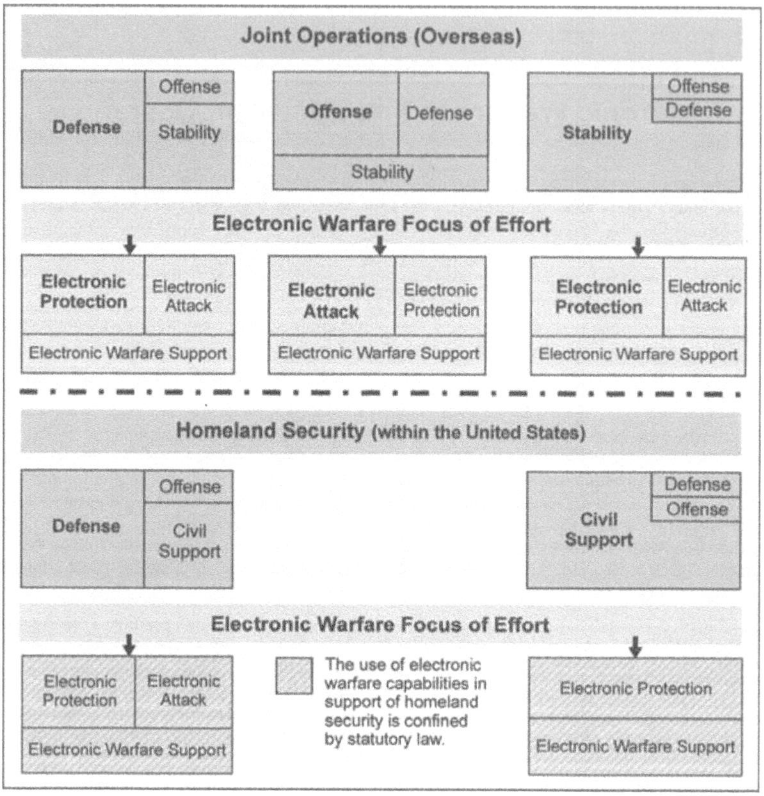

Figure 2-1. Electronic warfare weight of effort during operations

2-3. Full spectrum operations involve more than executing all elements of operations simultaneously. They require that commanders and staffs consider their unit's capabilities and capacities relative to each of the elements of full spectrum operations. Commanders consider how much can be accomplished simultaneously, how much can be phased, and what nonorganic resources may be available to solve problems. The same applies to EW in support of full spectrum operations. Commanders and staffs determine which resident and joint force EW capabilities to leverage in support of each element of full spectrum operations. Weighting the EW focus of effort within each of the elements assists commanders and their staffs in visualizing how EW capabilities can support their operations. Commanders combine offensive, defensive, and stability or civil support operations to seize, retain, and exploit the initiative. As they apply the appropriate level of EW effort to support these elements, commanders can seize, retain, and exploit the initiative within the electromagnetic environment.

THE APPLICATION OF ELECTRONIC WARFARE

2-4. To support full spectrum operations and achieve the goal of electromagnetic spectrum dominance, commanders fully integrate EW capabilities and apply them across the elements of combat power. Leadership and information are applied through, and multiply the effects of, the other six elements of combat power. Paragraphs 2-5 through 2-16 discuss the elements of combat power and how EW capabilities can support them.

IN SUPPORT OF LEADERSHIP

2-5. Leadership initiates the conditions for success. Commanders balance the ability to mass the effects of lethal and nonlethal systems with the requirements to deploy and sustain the units that employ those systems. Generating and maintaining combat power throughout an operation is essential. Today's operational environments require leaders who are competent, confident, and informed in using and protecting combat capabilities that operate within the electromagnetic spectrum. Commanders plan, prepare, execute, and assess EW operations to dominate the electromagnetic spectrum within their operational environment. To accomplish this domination, commanders effectively apply and integrate EW operations across the warfighting functions.

IN SUPPORT OF INFORMATION TASKS AND CAPABILITIES

2-6. Information is the element of combat power consisting of meaningful facts, data, and impressions used to develop a common situational understanding, to enable battle command, and to affect the operational environment. (See FM 3-0 for a discussion of combat power.) In modern conflict, gaining information superiority has become as important as lethal action in determining the outcome of operations. *Information superiority* is the operational advantage derived from the ability to collect, process, and disseminate an uninterrupted flow of information while exploiting or denying an adversary's ability to do the same (JP 3-13). To achieve this operational advantage, Army commanders direct efforts that contribute to information superiority. These efforts fall into four primary areas: Army information tasks; intelligence, surveillance, and reconnaissance; knowledge management; and information management. (See FM 3-0 for a discussion of information superiority.)

2-7. The Army information tasks are used to shape a commander's operational environment. These tasks are information engagement, command and control warfare, information protection, operations security, and military deception. Information capabilities can be used to produce both destructive and constructive effects. For example, destructive actions use information capabilities against the enemy's command and control system and other assets to reduce their combat capability. Constructive actions use information capabilities to inform or influence a particular audience or as a means to affect enemy morale. Although applicable to all elements of full spectrum operations, EW capabilities play a major role in enabling and supporting the execution of the command and control warfare and information protection tasks.

2-8. *Command and control warfare* is the integrated use of physical attack, electronic warfare, and computer network operations, supported by intelligence, to degrade, destroy, and exploit an enemy's or adversary's command and control system or to deny information to it (FM 3-0). It includes operations intended to degrade, destroy, and exploit an enemy's or adversary's ability to use the electromagnetic spectrum and computer and telecommunications networks. *Information protection* is active or passive measures that protect and defend friendly information and information systems to ensure timely, accurate, and relevant friendly information. Information protection denies enemies, adversaries, and others the opportunity to exploit friendly information and information systems for their own purposes (FM 3-0). Table 2-1 shows capabilities, intended effects, staff responsibilities, and functional cells for the command and control warfare and information protection tasks. (For further information on the information tasks, refer to FM 3-0.)

Table 2-1. Two Army information tasks: command and control warfare and information protection

<table>
<tr><th>Army Information Tasks</th><th>Capabilities</th><th>Staff Responsibility</th><th>Functional Coordinating Cell</th><th>Intended Effects</th><th>Integrating Process</th></tr>
<tr><td>Command and Control Warfare</td><td>Physical Attack
Electronic Attack
Computer Network Attack
Electronic Warfare Support
Computer Network Exploitation</td><td>G-3/G-2</td><td>Fires</td><td>Degrade, disrupt, destroy and exploit enemy command and control</td><td rowspan="2">Operations Process</td></tr>
<tr><td>Information Protection</td><td>Information Assurance
Computer Network Defense
Electronic Protection</td><td>G-6</td><td>Network Operations</td><td>Protect friendly computer networks and communication means</td></tr>
<tr><td colspan="6">G-2 assistant chief of staff, intelligence
G-3 assistant chief of staff, operations
G-6 assistant chief of staff, signal</td></tr>
</table>

2-9. To support these information tasks, commanders ensure EW is coordinated, integrated, and synchronized with all other tasks. This occurs within the operations process through the various functional and integrating cells. Table 2-2 illustrates EW capabilities, actions, and objectives that support the command and control warfare and information protection tasks.

Table 2-2. Electronic warfare support to two Army information tasks

Information Tasks	Command and Control Warfare	Information Protection
Electronic warfare supports by	Locating and identifying threat command and control systems. Denying, disrupting, degrading, and/or destroying the enemy's command and control system. Supporting and complementing computer network attack and computer network exploitation operations.	Deconflicting spectrum usage with the spectrum manager. Hardening equipment against electromagnetic interference. Emissions control.
Action	Electronic attack (jamming, antiradiation missiles). Directed energy and electromagnetic spectrum area denial systems. Expendables (chaff, decoys, and flares). Electronic warfare support/signals intelligence.	Frequency agility in radios. Electronic shielding for systems. Electronic masking. Processes to counter intrusion. Implementing emissions control procedures to safeguard friendly systems and facilities from the effects of friendly and enemy electronic attack.
Objective or Effect	Detect, deny, disrupt or degrade, and destroy.	Control and protection.

In Support of the Warfighting Functions

2-10. EW capabilities support each of the six warfighting functions. Examples of specific supporting capabilities are given in the following paragraphs.

Movement and Maneuver

2-11. The *movement and maneuver warfighting function* is the related tasks and systems that move forces to achieve a position of advantage in relation to the enemy. Direct fire is inherent in maneuver, as is close combat (FM 3-0). EW capabilities that enable the movement and maneuver of Army forces include—

- Suppression and destruction of enemy integrated air defenses.
- Denial of enemy information systems and intelligence, surveillance, and reconnaissance sensors.
- Target designation and range finding.
- Protection from effects of friendly and enemy EW.
- Lethal and nonlethal effects against enemy combat capability (personnel, facilities, and equipment).
- Threat warning and direction finding.
- Use of the electromagnetic spectrum to counter improvised explosive device operations.
- Electromagnetic spectrum obscuration, low observability, and multispectral stealth.

Intelligence

2-12. The *intelligence warfighting function* is the related tasks and systems that facilitate understanding of the operational environment, enemy, terrain, and civil considerations (FM 3-0). It includes tasks associated with intelligence, surveillance, and reconnaissance. EW capabilities that enable the intelligence warfighting function include—

- Increased access for intelligence collection assets (systems and personnel) by reducing antiaccess, antipersonnel, and antisystems threats.
- Increased capability to search for, intercept, identify, and locate sources of radiated electromagnetic energy in support of targeting, information tasks, and future operations.
- Increased capability in providing threat recognition and threat warning to the force.
- Indications and warning of threat emitters and radar.
- Denial and destruction of counter-intelligence, -surveillance, and -reconnaissance systems.

Fires

2-13. The *fires warfighting function* is the related tasks and systems that provide collective and coordinated use of Army indirect fires, joint fires, and command and control warfare, including nonlethal fires, through the targeting process (FM 3-0). It includes tasks associated with integrating command and control warfare. EW capabilities that enable the fires warfighting function include—

- Detection and location of targets radiating electromagnetic energy.
- Disruption, degradation, and destruction options for servicing targets. This includes information systems, targets requiring precision strike (such as minimal collateral damage and minimal weapons signature), hard and deeply buried targets, weapons of mass destruction, and power generation and infrastructure targets.
- Control, dispersion, or neutralization of combatant and noncombatant personnel with nonpersistent effects and minimum collateral damage (scalable and nonlethal).
- Area denial capabilities against vehicles, vessels, and aircraft.

Sustainment

2-14. The *sustainment warfighting function* is the related tasks and systems that provide support and services to ensure freedom of action, extend operational reach, and prolong endurance (FM 3-0). EW capabilities that enable the sustainment warfighting function include—

- Protection of sustainment forces from friendly and adversary use of EW in static or mobile environments.
- Enhanced electromagnetic environment situational awareness through the interception, detection, identification, and location of adversary electromagnetic emissions and by providing indications and warnings. (This information can assist in convoy planning, asset tracking, and targeting of potential threats to sustainment operations.)
- Countering improvised explosive devices to support ground lines of communication (includes counter-radio-controlled improvised-explosive-device systems and countering other threats triggered through the electromagnetic spectrum, such as lasers).
- Spectrum deconfliction and emissions control procedures in support of sustainment command and control.
- Electromagnetic spectrum obscuration, low-observability, and multispectral stealth (These capabilities provide protection during sustainment operations).

Command and Control

2-15. The *command and control warfighting function* is the related tasks and systems that support commanders in exercising authority and direction (FM 3-0). EW capabilities that enable the command and control warfighting function include—

- Protection of friendly critical information systems and command and control nodes, personnel, and facilities from the effects of friendly and adversary EW operations.
- Control of friendly EW systems through—
  - Frequency deconfliction.
  - Asset tracking.
  - Employment execution.
  - Reprogramming of EW systems.
  - Registration of all electromagnetic spectrum emitting devices with the spectrum manager (both prior to deployment and when new systems or devices are added to the deployed force).
- The development of EW command and control tools to enhance required coordination between Army and joint EW operations.
- EW operations integration, coordination, deconfliction, and synchronization through the EW working group (see chapter 3).
- Increased commander situational understanding through improved common operational picture input of electromagnetic spectrum- and EW-related information.
- EW operations monitoring and assessment.

Protection

2-16. The *protection warfighting function* is the related tasks and systems that preserve the force so the commander can apply maximum combat power (FM 3-0). EW capabilities and actions that enable the protection warfighting function include—

- Enhanced electromagnetic spectrum situational awareness through the interception, detection, identification, and location of adversary electromagnetic emissions used to providing indications and warnings of threat emitters and radars.
- Denial, disruption, or destruction of electromagnetic-spectrum-triggered improvised explosive devices and enemy air defense systems.
- Deception of enemy forces.
- Electromagnetic spectrum obscuration, low-observability, and multispectral stealth.
- EW countermeasures for platform survivability (air and ground).
- Area denial capabilities (lethal and nonlethal) against personnel, vehicles, and aircraft.
- Protection of friendly personnel, equipment, and facilities from friendly and enemy electronic attack, including friendly information systems and information. (This includes the coordination and use of both airborne and ground-based electronic attack with higher and adjacent units.)

SUMMARY

2-17. Army EW operations provide the land force commander capabilities to support full spectrum operations (offensive, defensive, and stability or civil support operations). EW supports full spectrum operations by applying EW capabilities to detect, deny, deceive, disrupt, or degrade and destroy enemy combat capability and by controlling and protecting friendly use of the electromagnetic spectrum. These capabilities—when applied across the warfighting functions—enable commanders to address a broad set of electromagnetic-spectrum-related targets to gain and maintain an advantage within the electromagnetic spectrum.

Chapter 3
Electronic Warfare Organization

A flexible organizational framework and capable, proficient electronic warfare personnel enable the commander's electronic warfare capability on the battlefield. This chapter discusses a framework that ensures coordination, synchronization, and integration of electronic warfare into full spectrum operations. This electronic warfare organizational framework supports current operations and is adaptable for future operations.

ORGANIZING ELECTRONIC WARFARE OPERATIONS

3-1. Operational challenges across the electromagnetic spectrum are expanding rapidly. As Army electronic warfare (EW) capabilities expand to meet these challenges, the organizational design required to coordinate, synchronize, integrate, and deconflict these capabilities must transform as rapidly. To meet current and future requirements, command and control of EW operations is built around the concept of EW working groups. Figure 3-1, page 3-2, illustrates the EW coordination organizational framework.

ARMY SERVICE COMPONENT COMMAND, CORPS, AND DIVISION LEVELS

3-2. A *working group* is a temporary grouping of predetermined staff representatives who meet to coordinate and provide recommendations for a particular purpose or function (FMI 5-0.1). The EW working group, when established, is responsible to the G-3 through the fires cell. An EW working group usually includes representation from the G-2, G-3, G-5, G-6, and G-7. (Joint doctrine calls this organization the EW coordination cell.) The EW working groups depicted in figure 3-1 (page 3-2) facilitate the internal (Army) and external (joint) integration, synchronization, and deconfliction of EW actions with fires, command and control, movement and maneuver, intelligence, sustainment and protection warfighting functions. Normally, EW working groups do not add additional structure to an existing organization. As depicted in figure 3-1, working groups vary in size and composition based on echelon.

3-3. Normally, the senior EW officer heads the EW working group and is accountable to the G-3 for integrating EW requirements. Working within the fires cell, the EW officer coordinates directly with the fire support coordinator for the integration of EW into the targeting process. This ensures EW capabilities are fully integrated with all other effects. Additional staff representation within EW working groups may include a fire support coordinator, a spectrum manager, a space operations officer, and liaison officers as required. Depending on the echelon, liaisons could include joint, interagency, and multinational representatives. When an Army headquarters serves as the headquarters of a joint task force or joint force land component command, the Army headquarters' working group becomes the joint force EW coordination cell.

3-4. When Army forces are employed as part of a joint or multinational force, they normally have EW representatives supporting higher headquarters' EW coordination organizations. These organizations may include the joint force commander's EW staff or the information operations cell within a joint task force. Sometimes a component EW organization may be designated as the joint EW coordination cell. (Chapter 6 discusses joint electronic warfare operations in more detail.) The overall structure of the combatant force and the level of EW to be conducted determine the structure of the joint EW coordination cell. The organization to accomplish the required EW coordination and functions varies by echelon.

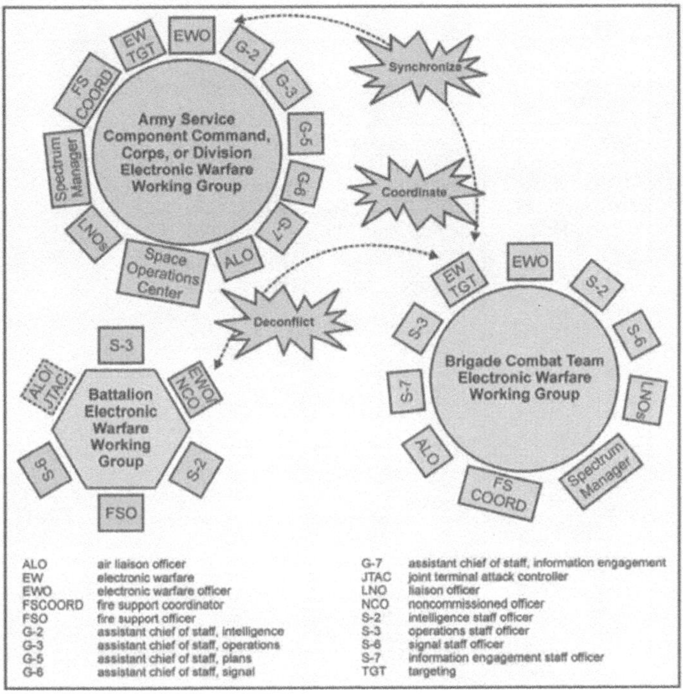

Figure 3-1. Electronic warfare coordination organizational framework

3-5. Regardless of the organizational framework employed, EW working groups perform specific tasks. Table 3-1 (page 3-3) details the functions of the EW working groups by echelon from battalion to Army Service component command. There is no formal organizational framework for EW at the company level (see paragraph 3-9).

Table 3-1. Functions of electronic warfare working groups

EW Working Group	Functions
Division and Above ALO EWO EW targeting G-2 G-3 G-5 G-6 G-7 FSCOORD LNOs Spectrum manager Space support officer	**Peacetime: Division—ASCC** • Conduct long-range electronic warfare planning in support of theater or combatant command requirements. • Integrate electronic warfare into operation plans and concept plans. • Develop electronic warfare supporting plans to operation plans and contingency plans. • Coordinate joint electronic warfare training and exercises. • Develop information and knowledge necessary to support contingency planning (for example, joint restricted frequency list development, spectrum management, and deconfliction). **Wartime: Division—ASCC** • Serve as the joint force land component or joint task force EW working group. • When directed, serve as the jamming control authority. • Develop and promulgate electronic warfare policies and support higher level policies. • Identify and coordinate intelligence support requirements for electronic warfare. • Plan, coordinate, and assess offensive and defensive electronic warfare requirements. • Plan, coordinate, synchronize, deconflict, and assess electronic warfare operations. • Maintain current assessment of electronic warfare resources available to the commander. • Prioritize electronic warfare effects and targets. • Predict effects of friendly and enemy electronic warfare. • Coordinate spectrum management and radio frequency deconfliction with G-6 and J-6. • Plan, assess, and implement friendly electronic security measures. • Plan, coordinate, integrate, and deconflict electronic warfare effects within the operations process.
Brigade S-3 EWO EW targeting FSCOORD S-2 S-6 ALO LNOs S-7 Spectrum manager	**Peacetime:** • Develop electronic warfare supporting requirements to operations plans and exercises. **Wartime:** • Support electronic warfare policies. • Plan, prepare, execute, and assess electronic warfare operations. • Integrate electronic warfare intelligence preparation of the battlefield into the operations process. • Identify and coordinate intelligence support requirements for BCT and subordinate units' electronic warfare operations. • Assess offensive and defensive electronic warfare requirements. • Maintain current assessment of electronic warfare resources available to unit. • Prioritize BCT and subordinate units' electronic warfare targets. • Plan, coordinate, and assess friendly electronic warfare operations. • Implement friendly electronic security measures (for example, electromagnetic spectrum mitigation and network protection). • When directed, serve as the jamming control authority.
Battalion EWO/NCO FSO S-2 S-3 S-6 JTAC	**Peacetime:** • Support BCT electronic warfare requirements to operations and exercises. **Wartime:** • Evaluate electronic warfare offensive, defensive, and support requirements. • Coordinate electronic warfare operations with higher headquarters. • Identify and coordinate intelligence support requirements with higher headquarters. • Execute electronic warfare in support of current operations. • Assess electronic warfare operations.

ALO	air liaison officer	G-7	assistant chief of staff, information engagement
ASCC	Army service component command	J-6	communications system directorate of a joint staff
BCT	brigade combat team	JTF	joint task force
EW	electronic warfare	JTAC	joint terminal attack controller
EWO	electronic warfare officer	LNO	liaison officer
FSCOORD	fire support coordinator	NCO	noncommissioned officer
G-2	assistant chief of staff, intelligence	S-2	intelligence staff officer
G-3	assistant chief of staff, operations	S-3	operations staff officer
G-4	assistant chief of staff, logistics	S-6	signal staff officer
G-5	assistant chief of staff, plans	S-7	information engagement staff officer
G-6	assistant chief of staff, signal		

Brigade Level

3-6. At the brigade level, the EW officer heads the EW working group and is accountable to the S-3 for integrating EW requirements. Additional staff representation within EW working groups at the brigade combat team level may include the fire support coordinator, EW targeting technician, S-2, S-6, spectrum manager, S-7, and liaison officers as required.

3-7. The EW working group at the brigade combat team coordinates with the higher echelon EW working groups. The brigade working group plays an important role in requesting and integrating joint air and ground EW support. It also manages the brigade's organic EW "fight" within the fires cell. The EW officer works as part of the brigade combat team staff. In this position, the EW officer synchronizes, integrates, and deconflicts brigade combat team EW actions with the EW working group at division level. Although EW falls under the control of the S-3, EW officers are fully immersed in fires targeting and planning to ensure proper use and coordination of EW. See table 3-1, page 3-3, for an outline of the functions of the brigade combat team EW working group.

Battalion Level

3-8. At the battalion level, the EW officer or noncommissioned officer leads the EW working group and is accountable to the S-3 for integrating EW requirements. Additional staff representation within EW working groups at the battalion level may include the S-2, S-6, fire support officer, and a joint terminal attack controller when assigned. The battalion EW working group coordinates battalion EW operations with the brigade combat team EW working group. See table 3-1, page 3-3, for an outline of the functions of the battalion EW working group.

Company Level

3-9. At the company level, trained EW personnel holding an additional skill identifier of 1K (tactical EW operations) or 1J (operational EW operations) perform several tasks. They advise the commander on the employment of EW equipment, track EW equipment status, assist operators in the use and maintenance of EW equipment, and coordinate with higher headquarters EW working groups.

PLANNING AND COORDINATING ELECTRONIC WARFARE ACTIVITIES

3-10. Key personnel involved in the planning and coordination of EW activities are—

- G-3 and S-3 staff.
- EW officer.
- Fire support coordinator.
- G-2 and S-2 staff.
- G-6 and S-6 staff.
- Electromagnetic spectrum manager.
- Liaisons.

G-3 or S-3 Staff

3-11. The G-3 or S-3 staff is responsible for the overall planning, coordination, and supervision of EW activities, except for intelligence. The EW officer is part of the G-3 or S-3 staff. The G-3 or S-3 staff—

- Plans for and incorporates EW into operation plans and orders, in particular within the fire support plan and the information operations plan (in joint operations).
- Tasks EW actions to assigned and attached units.
- Exercises control over electronic attack, including integration of electromagnetic deception plans.

- Directs electronic protection measures the unit will take based on recommendations from the G-6 or S-6, the EW officer, and the EW working group.
- Coordinates and synchronizes EW training with other unit training requirements.
- Coordinates and synchronizes EW training with other unit training requirements.
- Issues EW support tasks within the unit intelligence, surveillance, and reconnaissance plan. These tasks are according to the collection plan and the intelligence synchronization matrices developed by the G-2 or S-2 and the collection manager.
- Coordinates with the EW working group to ensure planned EW operations support the overall tactical plan.
- Integrates electronic attack as a form of fires within the fires cell.

ELECTRONIC WARFARE OFFICER

3-12. As a member of the G-3 or S-3 staff, the EW officer plans, coordinates, and supports the execution of EW. The EW officer—

- Leads the EW working group.
- Plans, coordinates, and assesses EW offensive, defensive, and support requirements.
- Supports the G-2 or S-2 during intelligence preparation of the battlefield.
- Supports the fire support coordinator to ensure electronic attack fires are integrated with all other effects.
- Plans, assesses, and implements friendly electronics security measures.
- Prioritizes EW effects and targets with the fire support coordinator.
- Plans and coordinates EW operations across functional and integrating cells.
- Deconflicts EW operations with the spectrum manager.
- Maintains a current assessment of available EW resources.
- Participates in other cells and working groups (as required) to ensure EW integration.
- Serves as EW subject matter expert on existing EW rules of engagement.
- When designated, serves as the jamming control authority.
- Prepares, submits for approval, and supervises the issuing and implementation of fragmentary orders for EW operations.

G-2 OR S-2 STAFF

3-13. The G-2 or S-2 staff advises the commander and staff on the intelligence aspects of EW. The G-2 or S-2 staff—

- Provides threat data to support programming of unit EW systems and deconfliction of their use by the EW working group.
- Ensures that electronic order of battle requirements are included in the intelligence collection plan.
- Determines enemy EW organizations, disposition, capabilities, and intentions via collection and analysis.
- Determines enemy EW vulnerabilities and high-value targets.
- Assesses effects of friendly EW operations on the enemy.
- Helps prepare the intelligence-related portion of the EW running estimate.
- Provides input to the restricted frequency list by recommending guarded frequencies.
- Provides updates on the rapid electronic order of battle.
- Maintains appropriate threat EW databases.
- Works with the EW working group to ensure that intelligence collection is synchronized with EW requirements and deconflicted with planned EW actions. Ensures that EW threat data is deconflicted with friendly electromagnetic spectrum needs.

Network Operations Officer

3-14. The network operations officer (in the G-6 or S-6 staff) coordinates the communications network for the following services:

- Preparing the electronic protection policy on behalf of the commander.
- Assisting in preparing EW plans and orders.
- Reporting all enemy electronic attack activity detected by friendly communications and electronics elements to the EW working group for counteraction.
- Assisting the unit EW officer with resolving EW systems maintenance and communications fratricide problems.

Spectrum Manager

3-15. The spectrum manager coordinates electromagnetic spectrum use for a wide variety of communications and electronic resources. The spectrum manager—

- Issues the signal operating instructions.
- Provides all spectrum resources to the task force.
- Coordinates for spectrum usage with higher echelon G-6 or S-6, and applicable host-nation and international agencies as necessary.
- Coordinates the preparation of the restricted frequency list and issuance of emissions control guidance.
- Coordinates frequency allotment, assignment, and use.
- Coordinates electromagnetic deception plans and operations in which assigned communications resources participate.
- Coordinates measures to reduce electromagnetic interference.
- Coordinates with higher echelon spectrum managers for electromagnetic interference resolution that cannot be resolved internally.
- Assists the EW officer in issuing guidance in the unit (including subordinate elements) regarding deconfliction and resolution of interference problems between EW systems and other friendly systems.
- Participates in the EW working group to deconflict friendly electromagnetic spectrum requirements with planned EW operations and intelligence collection.

SUMMARY

3-16. The organizational framework for EW coordination and functions varies by echelon. The necessity to form an EW working group is largely based on the overall structure of the combatant force and the level of EW to be conducted. During unified actions, other Service EW officers, signals intelligence officers, and EW asset representatives are invaluable to Army EW working groups in the planning, preparation, execution, and assessment of EW operations. As Army EW capabilities and concepts for employment continue to evolve, so do the organizational designs that ensure their effective command and control and execution in support of operations.

Chapter 4

Electronic Warfare and the Operations Process

The *operations process* consists of the major command and control activities performed during operations: planning, preparing, executing, and continuously assessing the operation. The commander drives the operations process (FM 3-0). These activities occur continuously throughout an operation, overlapping and recurring as required (see figure 4-1). The staff electronic warfare officer is actively involved in the operations process. Electronic warfare planning, preparation, execution, and assessment require collective expertise from operations, intelligence, signal, and battle command. The electronic warfare officer—through the unit's electronic warfare working group—integrates efforts across the warfighting functions. This ensures that electronic warfare operations support the commander's objectives.

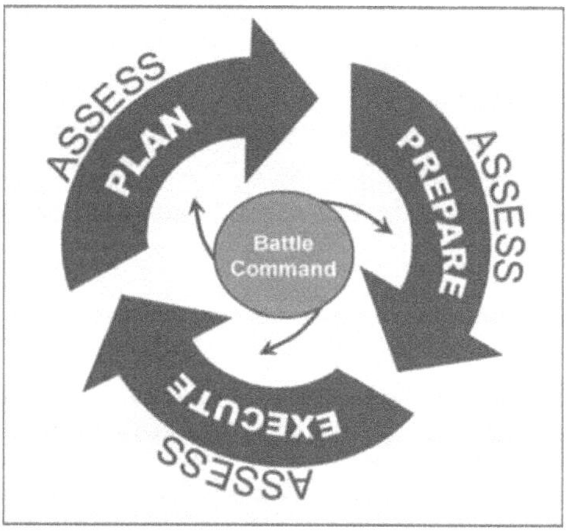

Figure 4-1. The operations process

SECTION I — ELECTRONIC WARFARE PLANNING

4-1. Electronic warfare (EW) planning is based on three main considerations. The first is applying the military decisionmaking process (MDMP). EW planners understand and follow its seven steps. In a time-constrained environment they still follow all seven steps, abbreviating the MDMP process appropriately. Additionally, EW planners apply EW integrating processes. They understand how EW actions contribute to operations. They integrate and synchronize EW activities starting with planning and continuing throughout operations. Finally, EW planners apply EW employment considerations according to the characteristics of EW capabilities.

THE MILITARY DECISIONMAKING PROCESS

4-2. EW planning minimizes fratricide and optimizes operational effectiveness during execution. Therefore, EW planning occurs concurrently with other operational planning during the MDMP. The MDMP synchronizes several processes, including intelligence preparation of the battlefield (IBP) (see FM 34-130), the targeting process (see FM 6-20-10), and risk management (see FM 5-19). These processes occur continuously during operations.

4-3. Depending on the organizational echelon, the staff EW officer leads EW planning through the EW working group. (The EW working group at echelons above brigade is sometimes referred to as an EW coordination cell.) An EW working group is normally supported by representatives from the G-2 or S-2, G-3 or S-3, G-6 or S-6, and other staff as required. Other staff representatives can include the fire support coordinator or fire support officer, spectrum manager, air liaison officer, space officer, and liaison officers. Paragraphs 4-5 through 4-33 outline key EW contributions to the processes and planning actions that occur during the seven steps of the MDMP. (FM 5-0 discusses the MDMP.)

RECEIPT OF MISSION

4-4. Commanders begin the MDMP upon receiving or anticipating a new mission. During this first step, commanders issue their initial guidance and initial information requirements or commander's critical information requirements.

4-5. Upon receipt of a mission, the staff EW officer alerts the staff members supporting the EW working group. The EW officer and support staff begin to gather the resources required for mission analysis. Resources might include a higher headquarters operation order or plan, maps of the area of operations, electronic databases, required field manuals and standing operating procedures, current running estimates, and reachback resources (see appendix F). The EW officer also provides input to the staff's initial assessment and updates the EW running estimate. As part of this update, the EW officer identifies all friendly EW assets and resources and their status. The EW officer also provides this information throughout the operations process. This includes monitoring, tracking, and seeking out information relating to EW operations to assist the commander and staff.

MISSION ANALYSIS

4-6. Planning includes a thorough mission analysis. Both the process and products of mission analysis help commanders refine their situational understanding and determine their restated mission. (See FM 5-0 for more details.) The EW officer and supporting members of the EW working group contribute to the overall mission analysis by participating in IPB and through the planning actions discussed in paragraphs 4-7 through 4-14. (Paragraphs 4-35 to 4-40 discuss EW input to IPB during operations.)

4-7. The EW officer and EW working group members—

- Convene the appropriate EW working group.
- Determine known facts, status, or conditions of forces capable of EW operations as defined in the commander's planning documents, such as a warning order or operation order.
- Identify EW planning support requirements and develop support requests as needed.

4-8. The EW officer and EW working group members support the G-2 and S-2 in IPB by—

- Determining the threat's dependence on the electromagnetic spectrum.
- Determining the threat's EW capability.

- Determining the threat's intelligence system collection capability.
- Determining which threat vulnerabilities relate to the electromagnetic spectrum.
- Determining how the operational environment affects EW operations using the operational variables and mission variables as appropriate.
- Initiating, refining, and validating information requirements and requests for information.

4-9. The EW officer and EW working group members—

- Determine facts and develop necessary assumptions relevant to EW such as the status of EW capability at probable execution and time available.
- Analyze the commander's mission and intent from an EW perspective.
- Identify constraints relevant to EW—
  - Actions EW operations must perform.
  - Actions EW operations cannot perform.
  - Other constraints.
- Analyze mission, enemy, terrain and weather, troops and support available, time available and civil considerations from the EW perspective.

4-10. The EW officer and EW working group members determine enemy and friendly centers of gravity and list their critical capabilities, requirements, and vulnerabilities from an EW perspective. (They determine how EW capabilities can best attack an enemy's command and control system.) The center of gravity analysis process outlined in figure 4-2 helps identify and list the critical vulnerabilities of enemy centers of gravity. The EW officer and EW working group members also list the critical requirements associated with the identified command and control critical capability (or command and control nodes) and then identify the critical vulnerabilities associated with the critical requirements. Through this process, the EW officer and EW working group members help determine which vulnerabilities can be engaged by EW capabilities to produce a decisive outcome.

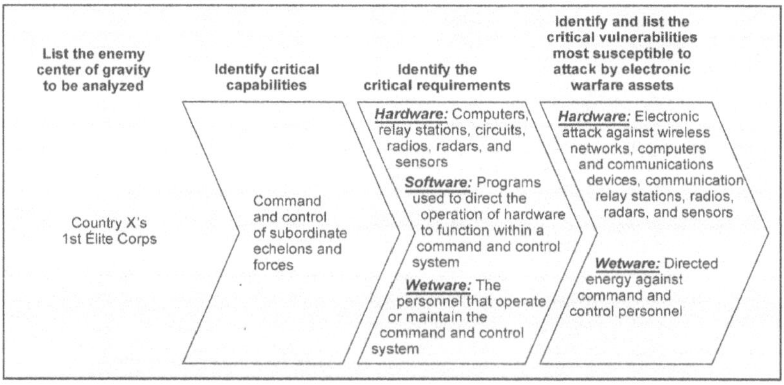

Figure 4-2. Example of analysis for an enemy center of gravity

4-11. Additionally, the EW officer and EW working group members determine how EW can help protect friendly centers of gravity. The center of gravity analysis process as outlined in figure 4-2 can also be used help identify critical vulnerabilities of friendly centers of gravity. The EW officer and EW working group members list the critical requirements associated with the identified friendly command and control critical capability. Then, the EW officer and EW working group members identify the critical vulnerabilities associated with the critical requirements. These vulnerabilities can help determine how to best use EW capabilities to defend or protect friendly centers of gravity from enemy attack. Key to this portion of the analysis is to assess the potential impact of EW operations on friendly information systems such as electromagnetic interference.

4-12. The EW officer and EW working group members identify and list—

- High-value targets that can be engaged by EW capabilities.
- Tasks that EW forces perform according to EW subdivision (electronic attack, electronic warfare support, and electronic protection) in support of the warfighting functions. These include—
  - Determining specified EW tasks.
  - Determining implied EW tasks.

4-13. The EW officer and EW working group members—

- Conduct initial EW force structure analysis to determine if sufficient assets are available to perform the identified EW tasks. (If organic assets are insufficient, they draft requests for support and augmentation.)
- Conduct an initial EW risk assessment and review the risk assessment done by the entire working group.
- Provide EW perspective in the development of the commander's restated mission.
- Assist in development of the mission analysis briefing for the commander.

4-14. By the conclusion of mission analysis, the EW officer and EW working group members generate or gather the following products and information:

- The initial information requirements for EW operations.
- A rudimentary command and control nodal analysis of the enemy.
- The list of EW tasks required to support the mission.
- A list of assumptions and constraints related to EW operations.
- The planning guidance for EW operations.
- EW personnel augmentation or support requirements.
- An update of the EW running estimate.
- EW portion or input to the commander's restated mission.

Course of Action Development

4-15. After receiving the restated mission, commander's intent, and commander's planning guidance, the staff develops courses of action (COAs) for the commander's approval. Figure 4-3 depicts the required input to COA development and identifies the key contributions made by the EW officer and EW working group members during the process and output stages (center and right of figure 4-3). The actions the EW officer and EW working group members perform to support COA development are discussed in more detail in paragraphs 4-16 through 4-20.

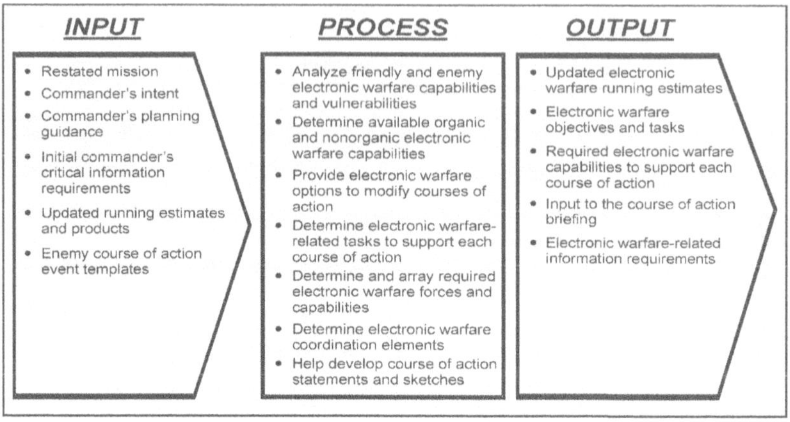

Figure 4-3. Course of action development

4-16. The EW officer and EW working group members contribute to COA development through the following planning actions—

- Determining which friendly EW capabilities are available to support the operation, including organic and nonorganic capabilities for planning.
- Determining possible friendly and enemy EW operations, including identifying friendly and enemy vulnerabilities.

4-17. Additionally, the EW officer and EW working group members help develop initial COA options by—

- Identifying COA options that may be feasible based on their functional expertise (while brainstorming of COAs).
- Providing options to modify a COA to enable accomplishing a requirement within the EW area of expertise.
- Identifying information (relating to EW options) that may impact other functional areas and sharing that information immediately.
- Identifying the EW-related tasks required to support the COA options.

4-18. The EW officer and EW working group members determine the forces required for mission accomplishment by—

- Determining the EW tasks that support each COA and how to perform those tasks based on available forces and capabilities. (Available special technical operations capabilities are considered in this analysis.)
- Providing input and support to proposed deception options.
- Ensuring the EW options provided in support of all possible COAs meet the established screening criteria.

4-19. The EW officer and EW working group members identify EW supporting tasks and their purpose in supporting any decisive, shaping, and sustaining operations as each COA is developed. These EW tasks include those—

- Focused on defeating the enemy.
- Required to protect friendly force operations.

4-20. The EW officer and EW working group members assist in developing the COA briefing as required. By the conclusion of COA development, the EW officer and EW working group members generate or gather the following products and information:

- A list of EW objectives and desired effects related to the EW tasks.
- A list of EW capabilities required to perform the stated EW tasks for each COA.
- The information and intelligence requirements for performing the EW tasks in support of each COA.
- An update to the EW running estimate.

Course of Action Analysis (War-gaming)

4-21. The COA analysis allows the staff to synchronize the elements of combat power for each COA and to identify the COA that best accomplishes the mission. It helps the commander and staff to—

- Determine how to maximize the effects of combat power while protecting friendly forces and minimizing collateral damage.
- Further develop a visualization of the battle.
- Anticipate battlefield events.
- Determine conditions and resources required for success.
- Determine when and where to apply force capabilities.
- Focus IPB on enemy strengths and weaknesses as well as the desired end state.
- Identify coordination needed to produce synchronized results.
- Determine the most flexible COA.

Paragraphs 4-22 to 4-23 discuss actions the EW officer and EW working group members perform to support COA analysis. (See FM 5-0 for more information on war-gaming.)

4-22. During COA analysis, the EW officer and EW working group members synchronize EW actions and assist the staff in integrating EW capabilities into each COA. The EW officer and EW working group members address how each EW capability supports each COA. They apply these capabilities to associated time lines, critical events, and decision points in the synchronization matrix (see table 4-1). During this planning phase, the EW officer and EW working group members aim to—

- Analyze each COA from an EW functional perspective.
- Recommend any EW task organization adjustments.
- Identify key EW decision points.
- Provide EW data for synchronization matrix.
- Recommend EW priority intelligence requirements.
- Identify EW supporting tasks to any branches and sequels.
- Identify potential EW high-value targets.
- Assess EW risks created by telegraphing intentions, allowing time for enemy to mitigate effects, unintended effects of electronic attack, and the impact of asset or capability shortfalls.

4-23. By the conclusion of COA analysis (war-gaming), the EW officer and EW working group members generate or gather the following products and information:

- The EW data for the synchronization matrix.
- The EW portion of the branches and sequels.
- A list of high-value targets related to EW.
- A list of commander's critical information requirements related to EW.
- The risk assessment for EW operations in support of each COA.
- An update to the EW running estimate.

Table 4-1. Sample input to synchronization matrix

	TIME/EVENT	H - 8	H - hour	H + 8
	Enemy Actions	Enemy monitors movements	Defends from Security Zone	Commits reserve
	Decision Points	Launch deep attack		
MANEUVER	1st Brigade	Move on route Paula	Cross line of departure	Seize objective Nick
MANEUVER	2d Brigade	Move on route Mike	Cross line of departure	Seize objective Dave
MANEUVER	3rd Brigade	Move on route Sean		Forward passage of lines with 1st Brigade
MANEUVER	Aviation Brigade	Deep attack on objective Rose at H - 1		
MANEUVER	Division Cavalry		Screen northern flank	
FIRES	Fires Brigade	Preparation fires initiated at H - 5		
FIRES	Air Defense	Weapons hold	Weapons tight	Weapons tight
FIRES	C2W - EA - CNA - Physical Attack - CNE - ES	- Locate enemy ISR on maneuver routes - Deny and disrupt enemy ISR of maneuver routes at H - .5 to H - hour - Disrupt and destroy known enemy C2 nodes and IADS	- Activate CREW systems - Jamming (to disrupt/deny enemy C2 nodes) - Electronic deception - Provide indications and warnings to maneuver brigades	Disrupt and destroy enemy C2 system
C2			Tactical CP with lead brigade	

C2	command and control	EA	electronic attack
C2W	command and control warfare	ES	electronic warfare support
CNA	computer network attack	EW	electronic warfare
CNE	computer network exploitation	H-hour	specific time an operation or exercise begins
CP	command post	IADS	integrated air defense system
CREW	counter radio-controlled improvised explosive device electronic warfare	ISR	intelligence, surveillance, and reconnaissance

Note: This is not complete. Its intent is to show how EW can be integrated into a synchronization matrix.

COURSE OF ACTION COMPARISON

4-24. COA comparison starts with all staff members analyzing and evaluating the advantages and disadvantages of each COA from their perspectives. Staff members present their findings for the others' consideration. Using the evaluation criteria developed during COA analysis, the staff outlines each COA, highlighting its advantages and disadvantages. Comparing the strengths and weaknesses of the COAs identifies their advantages and disadvantages with respect to each other. (See FM 5-0 for further discussion of COA comparison).

4-25. During COA comparison, the EW officer and EW working group members compare COAs based on the EW-related advantages and disadvantages (see center of figure 4-4). Typically, planners use a matrix to assist in the COA comparisons. The EW officer may develop an EW functional matrix to compare the COAs or to use the decision matrix developed by the staff. Regardless of the matrix used, the evaluation criteria developed prior to war-gaming are used to compare the COAs. Normally, the chief of staff or executive officer weights each criterion used for the evaluation based on its relative importance and the commander's guidance. (See FM 5-0 for more information on COA comparison and a sample decision matrix.)

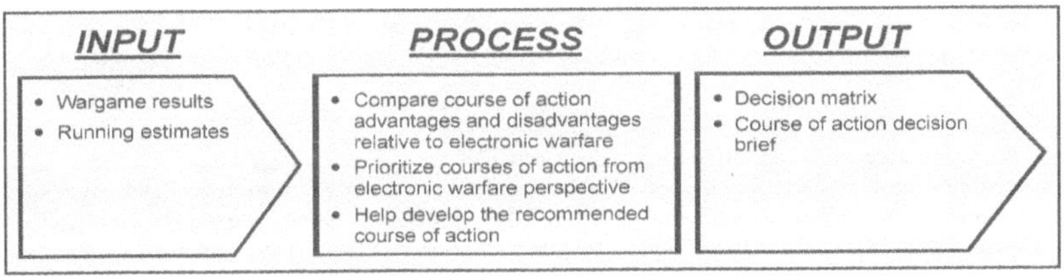

Figure 4-4. Course of action comparison

4-26. By the conclusion of COA comparison, the EW officer and EW working group members generate or gather the following products and information:

- A list of the pros and cons for each COA relative to EW.
- A prioritized list of the COAs from an EW perspective.
- An update to the EW running estimate if required.

COURSE OF ACTION APPROVAL

4-27. The COA approval process has three components. First, the staff recommends a COA, usually in a decision briefing. Second, the commander decides which COA to approve. Lastly, the commander issues the final planning guidance.

4-28. During COA approval, the EW officer supports the development of the COA decision briefing and the development of the warning order as required. If possible, the EW officer attends the COA decision briefing to receive the commander's final planning guidance. If unable to attend the briefing, the EW officer receives the final planning guidance from the G-3 or S-3. The final planning guidance is critical in that it normally provides—

- Refined commander's intent.
- New commander's critical information requirements to support the execution of the chosen COA.
- Risk acceptance.
- Guidance on priorities for the elements of combat power, orders preparation, rehearsal, and preparation.

4-29. After the COA decision has been made, the EW officer and EW working group members generate or gather the following products and information:

- An updated command and control nodal analysis of the enemy relevant to the selected COA.
- Required requests for information to refine the enemy command and control nodal architecture.
- Latest electronic order of battle tailored to the selected COA.
- Any new direction provided in the refined commander's intent.
- A list of any new commander's critical information requirements that can be used in support of EW operations.
- The warning order to assist developing EW operations required to support the operation order or plan.
- Refined input to the initial intelligence, surveillance, and reconnaissance (ISR) plan, including—
  - Any additional specific EW information requirements.
  - Updated potential collection assets for the unit's ISR plan.

Orders Production

4-30. Orders production consists of the staff preparing the operation order or plan by converting the selected COA into a clear, concise concept of operations. The staff also provides supporting information that enables subordinates to execute and implement risk controls. They do this by coordinating and integrating risk controls into the appropriate paragraphs and graphics of the order.

4-31. During orders production, the EW officer provides the EW operations input for several sections of the operation order or plan. See appendix B for the primary areas for EW operations input within an Army order or plan. The primary areas for EW input in a joint order, if required, also are shown in appendix B. (See CJCSM 3122.03C for the Joint Operation Planning and Execution System format).

DECISIONMAKING IN A TIME-CONSTRAINED ENVIRONMENT

4-32. In a time-constrained environment, the staff might not be able to conduct a detailed MDMP. The staff may choose to abbreviate the process as described in FM 5-0. The abbreviated process uses all seven steps of the MDMP in a shortened and less detailed manner.

4-33. The EW officer and core members of the EW working group meet as a regular part of the unit battle rhythm. However, the EW officer calls unscheduled meetings if situations arise that require time-sensitive planning. Regardless of how much they abbreviate the planning process, the EW officer and supporting members of the EW working group always—

- Update the EW running estimate in terms of assets and capabilities available.
- Update essential EW tasks with the requirements of the commander's intent.
- Coordinate support requests and intelligence requirements with appropriate staff elements and outside agencies.
- Provide EW input to fragmentary orders through the G-3 or S-3 as necessary to drive timely and effective EW operations.
- Deconflict planned EW actions with other uses of the spectrum, such as communications.
- Synchronize electronic attack and EW support actions.
- Synchronize other intelligence collection in support of EW requirements.
- Deconflict EW activities specifically with aviation operations.
- Synchronize EW support to the command and control warfare and information protection information tasks.

THE INTEGRATING PROCESSES AND CONTINUING ACTIVITIES

4-34. Commanders use several integrating processes and continuing activities to synchronize operations throughout the operations process. (See figure 4-5.) The EW officer ensures EW operations are fully synchronized and integrated within these processes and continuing activities. Other staff members supporting the EW working group assist the EW officer. Paragraphs 4-35 through 4-52 outline some key integrating processes and continuing activities. These processes and activities require EW officer involvement throughout the operations process.

Plan	*Prepare*	*Execute*
	Assess	

Integrating Processes →
- Intelligence preparation of the battlefield
- Targeting
- Intelligence, surveillance, and reconnaissance synchronization
- Composite risk management
- Knowledge management

Continuing Activities →
- Intelligence, surveillance, and reconnaissance
- Security operations
- Protection
- Liaison and coordination
- Terrain management
- Information management
- Airspace command and control

Source: FM 3-0

Figure 4-5. Integrating processes and continuing activities

INTELLIGENCE PREPARATION OF THE BATTLEFIELD

4-35. *Intelligence preparation of the battlefield* is the systematic, continuous process of analyzing the threat and environment in a specific geographic area. Intelligence preparation of the battlefield is designed to support the staff estimate and military decisionmaking process. Most intelligence requirements are generated as a result of the intelligence preparation of the battlefield process and its interrelation with the decisionmaking process (FM 34-130). The G-2 or S-2 leads IPB planning with participation by the entire staff. This planning activity is used to define and understand the operational environment and the options it presents to friendly and adversary forces. Only one IPB planning activity exists within each headquarters; all affected staff cells participate. (FM 2-0 provides more information on IPB.) Paragraphs 4-36 through 4-40 discuss how the EW officer and the EW working group support IPB during operations.

4-36. In addition to the input provided to the initial IPB (during step 2 of mission analysis), the EW officer supports IPB throughout the operations process by providing input related to EW operations. (See figure 4-6.) This input includes (but is not limited to) the following EW considerations:

- Evaluating the operational environment from an EW perspective.
- Describing how the effects of the operational environment may impact EW operations.
- Evaluating the threat's capabilities; doctrinal principles; and tactics, techniques, and procedures from an EW perspective.
- Determining threat COAs.

4-37. When evaluating the operational environment from an EW perspective, the EW officer—

- Determines the electromagnetic environment within the defined physical environment:
  - Area of operations.
  - Area of influence.
  - Area of interest.
- Uses electronic databases to identify gaps.
- Identifies adversary fixed EW sites such as EW support and electronic attack sites.
- Identifies airfields and installations that support, operate, or house adversary EW capabilities.
- In coordination with the G-2 or S-2 and G-6 or S-6, helps identify enemy electromagnetic spectrum usage and requirements within the area of operations and area of interest.

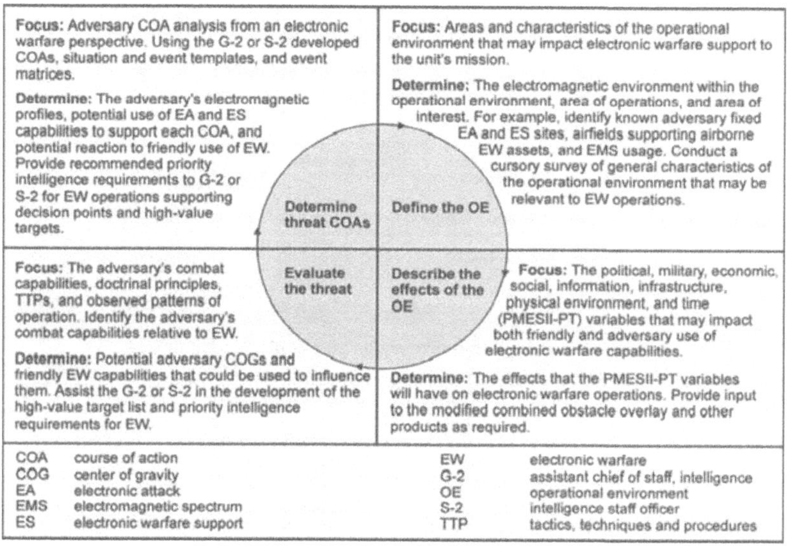

Figure 4-6. Electronic warfare support to intelligence preparation of the battlefield

4-38. When describing how the variables of the operational environment may impact EW operations, the EW officer—

- Focuses on characteristics of both the land and air domains using the factors of observation and fields of fire, avenues of approach, key and decisive terrain, obstacles, and cover and concealment.
- Identifies key terrain that may provide protection for communications and target acquisition systems from exploitation or disruption.
- Identifies how terrain affects line of sight, including effects on both communications and non-communications emitters.
- Evaluates how vegetation affects radio wave absorption and antenna height requirements.
- Locates power lines and their potential to interfere with radio waves.
- Assesses most likely and most dangerous avenues of approach (air, ground) and where EW operations would likely be positioned to support these approaches.

- If operating within urban terrain, considers how the infrastructure—power plants, power grids, structural heights, and communications and media nodes—may restrict or limit EW capabilities.
- Assists the G-2 or S-2 with the development of a modified combined obstacle overlay.
- Determines how weather—visibility, cloud cover, rain, and wind—may affect ground-based and airborne EW operations and capabilities (for example, no-go weather conditions at an airborne EW launch and recovery base).
- Considers all other relevant aspects of the operational environment that affect EW operations, using the operational variables (PMESII-PT—political, military, economic, social, information, infrastructure, physical environment, and time) and mission variables (METT-TC—mission, enemy, terrain and weather, troops and support available, time available, and civil considerations).

4-39. When evaluating enemy capabilities, the EW officer and supporting staff examine doctrinal principles; tactics, techniques and procedures; and observed patterns of operation from an EW perspective. The EW officer—

- Uses the operational variables (PMESII-PT) and mission variables (METT-TC) to help determine the adversary's critical nodes.
- Collects the required data—operational net assessments, electronic order of battle, and electronic databases—to template the command and control critical nodes and the systems required to support and maintain them.
- Assists the G-2 in determining the adversary's EW-related threat characteristics (order of battle) by identifying—
  - Types of communications equipment available.
  - Types of noncommunications emitters.
  - Surveillance and target acquisition assets.
  - Technological sophistication of the threat.
  - Communications network structure.
  - Frequency allocation techniques.
  - Operation schedules.
  - Station identification methods.
  - Measurable characteristics of communications and noncommunications equipment.
  - Command, control, and communications structure of the threat.
  - Tactics from a communication perspective. Examples are how the enemy deploys command, control, and communications assets; whether or not communications systems are remote; and the level of discipline in procedures, communications security, and operations security.
  - Electronic deception capabilities.
  - Reliance on active or passive surveillance systems
  - Electromagnetic profiles of each node.
  - Unique electromagnetic spectrum signatures.
- Assists the G-2 or S-2 in center of gravity analysis. Helps identify the critical system nodes of the center of gravity and determines what aspects of the system should be engaged, exploited, or attacked to modify the system's behavior or to achieve a desired effect.
- Identifies organic and nonorganic EW capabilities available to achieve desired effects on identified high-value targets.
- Submits initial EW-related requests for information that describe the intelligence support required to support EW operations.
- Obtains the high-value target list, threat templates, and initial priority intelligence requirements list to assist in follow-on EW planning.

4-40. When determining adversary COAs, the EW officer—

- Assists the G-2 or S-2 in development of adversary COAs.
- Provides EW input to the situation templates.
- Ensures event templates include EW named areas of interests.
- Assists in providing EW options for target areas of interest.
- Assists in providing EW options to support decision points.
- Provides EW input to the event template and event matrix.

TARGETING

4-41. *Targeting* is the process of selecting and prioritizing targets and matching the appropriate response to them, considering operational requirements and capabilities (JP 3-0). A decide, detect, deliver, and assess methodology is used to direct friendly forces to attack the right target with the right asset at the right time. (See figure 4-7.) Targeting provides an effective method to match the friendly force capabilities against targets. Commander's intent plays a critical role in the targeting process. The targeting working group strives to thoroughly understand the commander's intent to ensure the commander's intended effects on targets are achieved.

4-42. An important part of targeting is identifying potential fratricide situations and performing the coordination measures to manage and control the targeting effort positively. The targeting working group and staff incorporate these measures into the coordinating instructions and appropriate annexes of the operation plans and orders. (FM 6-20-10 has more information on targeting.)

DECIDE		DETECT	
Determine	**Based on**	**Determine**	**Based on**
What (task): Enemy focused. Determine what EW tasks are essential to the success of the operation (enemy formation or function to influence, and desired targeting effect). Why (purpose): Friendly focused. Determine the purpose for the use of EA fires (for example, to clear transit routes for maneuvering forces).	• Commanders initial planning guidance • Mission analysis - Specified and implied tasks - Intelligence preparation of the battlefield + target value analysis = enemy courses of action and high-value targets • Commander's intent	Who/Where: Focused on detection. Assets are deployed to detect high-payoff targets. ES collection assets identify and locate targets that can be influenced by EA. Once targets are identified, EA fires can be used to influence the targets based on the identified weaknesses by the target assessment and SIGINT teams.	• COA development • Concept of fires • War-gaming • COA decision • Scheme of fires • High-payoff target list and attack guidance matrix • Reconnaissance and surveillance plan
ASSESS		**DELIVER**	
Determine	**Based on**	**Determine**	**Based on**
Effect: Identifies whether the intended effect achieved by the EA fires was successful or not.	• Operations plans or orders • EW task execution • Effects of EA fires • BDA • MOE • Target assessment teams • SIGINT team assessment	Who/When: Focused on delivery. Addresses the who and when portion of task (such as the jamming of a designated target and the duration desired).	• COA development • Concept of fires • War-gaming • COA decision • Scheme of fires • High-payoff target list and attack guidance matrix • Reconnaissance and surveillance plan

BDA battle damage assessment
COA course of action
EA electronic attack
ES electronic warfare support
EW electronic warfare
MOE measures of effectiveness
SIGINT signals intelligence

Figure 4-7. Electronic warfare in the targeting process

4-43. The EW officer thoroughly integrates electronic attack in the targeting process and integrates electronic attack fires into all appropriate portions of the operation plan, operation order, and other planning products. In support of EW targeting, the EW officer—

- Helps the targeting working group determine electronic attack requirements against specific high-payoff targets and high-value targets.
- Ensures electronic attack can meet the desired effect (in terms of the targeting objective).
- Coordinates with the signals intelligence staff element through the collection manager to satisfy EW support and electronic attack information requirements.
- Prepares the EW tab and the EW portion of the command and control warfare tab to the fires appendix.
- Provides electronic attack mission management through the tactical operations center or joint operations center and the tactical air control party (for airborne electronic attack).
- Provides electronic attack mission management as the jamming control authority for ground or airborne electronic attack when designated.
- Prepares and coordinates the EW annex for operation plans and operation orders.
- Determines and requests theater Army electronic attack support.
- Recommends to the G-3 or S-3 and the fire support coordinator or fire support officer whether to engage a target with electronic attack.
- Expedites electromagnetic interference reports to the targeting working group. (See appendix D for information on electromagnetic interference reporting.)

Decide

4-44. Decide is the first step in the targeting process. This step provides the overall focus for fires, a targeting plan, and some of the priorities for intelligence collection. As part of the staff in the fires cell, the EW officer assists the targeting working group in planning the target priorities for each phase and critical events of the operation. Initially, the targeting working group does not develop electronic attack targets using any special technique or separately from targets for physical destruction. However, as the process continues, these targets are passed through intelligence organizations and further planned using ISR procedures. The planned use of electronic attack is integrated into the standard targeting products (graphic or text-based). Products that involve electronic attack planning may include—

- High-payoff target list.
- Attack guidance matrix.
- Appendix 4 (Electronic Warfare) to Annex P (Information Operations) of the operation order. (At the time this manual was written, this was the current doctrine for operation orders. This appendix will be revised upon publication of the revised FM 5-0.)

Detect

4-45. Based on what the targeting working group identified as high-payoff targets during the decide step, collection assets are then deployed to detect them. The intelligence enterprise pairs assets to targets based on the collection plan and the current threat situation. When conducting electronic attack operations in support of command and control warfare, ISR units perform EW support tasks linked to and working closely with the electronic attack missions. Electronic warfare support units (with support from the target assessment and signals intelligence staff elements) provide the data—location, signal strength, and frequency of the target—to focus electronic attack assets on the intended target. These assets also identify the command and control system vulnerabilities open to attack by electronic attack assets.

Deliver

4-46. Once friendly force capabilities identify, locate, and track the high-payoff targets, the next step in the process is to deliver fires against those targets. Electronic attack assets must satisfy the attack guidance developed during the decide step. Close coordination between those conducting EW support and electronic attack is critical during the engagement. The EW officer facilitates this coordination and ensures electronic attack fires are fully synchronized and deconflicted with other fires. The EW officer remains aware of the potential for unintended effects between adjacent units when conducting electronic attack. The EW officer continually coordinates with adjacent unit EW officers to mitigate and deconflict these effects during cross-boundary operations. Normally, the G-3, S-3, or fire support coordinator provides requirements and guidance for this coordination and synchronization in the attack guidance matrix, intelligence synchronization matrix, spectrum management plan, and the EW input to the operation plan or operation order annexes and appendixes.

Assess

4-47. Once the target as been engaged, the next step is to assess the engagement's effectiveness. This is done through combat assessment, which involves determining the effectiveness of force employment during military operations. It consists of three elements:

- Munitions effects assessment.
- Battle damage assessment.
- Re-attack recommendations.

4-48. The first two elements, munitions effects assessment and battle damage assessment, are used to inform the commander on the effects achieved against targets and target sets. From this information, the G-2 or S-2 continues to analyze the threat's ability to further conduct and sustain combat operations (sometimes articulated in terms of the effects achieved against the threat's centers of gravity). The last element involves the assessment and recommendation whether or not to re-attack the targets.

4-49. The assessment of a jamming mission used against an enemy's command and control system is unlike fires that can be observed visually. The signals intelligence staff element and units executing the electronic attack mission coordinate continuously to assess mission effectiveness. Close coordination between sensor and shooter allows instant feedback on the success or failure of the intended jamming effects. It also can quickly provide the necessary adjustments to produce desired effects.

Intelligence, Surveillance, and Reconnaissance Synchronization

4-50. *Intelligence, surveillance, and reconnaissance synchronization* is the task that accomplishes the following: analyzes information requirements and intelligence gaps; evaluates available assets internal and external to the organization; determines gaps in the use of those assets; recommends intelligence, surveillance, and reconnaissance assets controlled by the organization to collect on the commander's critical information requirements; and submits requests for information for adjacent and higher collection support (FM 3-0). ISR synchronization considers all assets—both internal and external to the organization. It identifies information gaps and the most appropriate assets for collecting information to fill them.

4-51. Planning for ISR operations begins during mission analysis. Although led by the G-3 or S-3, it is supported by the entire staff, subordinate units, and external partners. ISR operations collect, process, store, display, and disseminate information from a multitude of collection sources. The staff thoroughly understands, integrates, and synchronizes the ISR plan across all echelons.

4-52. The EW officer ensures the ISR plan supports the EW-related information requirements determined during the planning process. The EW officer coordinates these requirements with the signals intelligence staff element through the G-2 or S-2.

EMPLOYMENT CONSIDERATIONS

4-53. EW has specific ground-based, airborne, and functional (electronic attack, electronic warfare support, or electronic protection) employment considerations. The EW officer ensures EW-related employment considerations are properly articulated early in the operations process. Each capability employed has certain advantages and disadvantages. The staff plans for all of these before executing EW operations.

Ground-Based Electronic Warfare Considerations

4-54. Ground-based EW capabilities support the commander's scheme of maneuver. Ground-based EW equipment can be employed by a dismounted Soldier or on highly mobile platforms. Due to the short-range nature of tactical signals direction finding, electronic attack assets are normally located in the forward areas of the battlefield, with or near forward units.

4-55. Ground-based EW capabilities have certain advantages. They provide direct support to maneuver units (for example, through counter-radio-controlled improvised-explosive-device EW and communications or sensor jamming). Ground-based EW capabilities support continuous operations and respond quickly to EW requirements of the ground commander. However, to maximize the effectiveness of ground-based EW capabilities, maneuver units must protect EW assets from enemy ground and aviation threats. EW equipment should be as survivable and mobile as the force it supports. Maneuver units must logistically support the EW assets, and supported commanders must clearly identify EW requirements.

4-56. Ground-based EW capabilities have certain limitations. They are vulnerable to enemy attack and can be masked by terrain. They are vulnerable to enemy electromagnetic deceptive measures and electronic protection actions. In addition, they have distance or propagation limitations against enemy electronic systems.

Airborne Electronic Warfare Considerations

4-57. While ground-based and airborne EW planning and execution are similar, they significantly differ in their EW employment time. Airborne EW operations are conducted at much higher speeds and generally have a shorter duration than ground-based operations. Therefore, the timing of airborne EW support requires detailed planning.

4-58. Airborne EW requires the following:

- A clear understanding of the supported commander's EW objectives.
- Detailed planning and integration.
- Ground support facilities.
- Liaisons between the aircrews of the aircraft providing the EW support and the aircrews or ground forces being supported.
- Protection from enemy aircraft and air defense systems.

4-59. Airborne EW capabilities have certain advantages. They can provide direct support to other tactical aviation missions such as suppression of enemy air defenses, destruction of enemy air defenses, and employment of high-speed antiradiation missiles. They can provide extended range over ground-based assets. Airborne EW capabilities can provide greater mobility and flexibility than ground-based assets. In addition, they can support ground-based units in beyond line-of-sight operations.

4-60. The limitations associated with airborne EW capabilities are time-on-station considerations, vulnerability to enemy electronic protection actions, electromagnetic deception techniques, and limited assets (support from nonorganic EW platforms need to be requested).

Electronic Attack Considerations

4-61. Electronic attack includes both offensive and defensive activities. (Chapter 1 provides a full definition of electronic attack). These activities differ in their purpose. Defensive electronic attack protects friendly personnel and equipment or platforms. Offensive electronic attack denies, disrupts, or destroys enemy capability. In either case, certain considerations are involved in planning for employing electronic attack:

- Friendly communications.
- Intelligence collection.
- Other effects.
- Nonhostile local electromagnetic spectrum use.

- Hostile intelligence collection.
- Persistency of effect.

4-62. The EW officer, the G-2 or S-2, the G-3 or S-3, the G-6 or S-6, the spectrum manager, and the G-7 or S-7 coordinate closely to avoid friendly communications interference that can occur when using EW systems on the battlefield. Coordination ensures that electronic attack systems frequencies are properly deconflicted with friendly communications and intelligence systems or that ground maneuver and friendly information tasks are modified accordingly.

4-63. The number of information systems, EW systems, and sensors operating simultaneously on the battlefield makes deconfliction with communications systems a challenge. The EW officer, the G-2 or S-2, the G-6 or S-6, and the spectrum manager plan and rehearse deconfliction procedures to quickly adjust their use of EW or communications systems.

4-64. Electronic attack operations depend on EW support and signals intelligence to provide targeting information and battle damage assessment. However, EW officers must keep in mind that not all intelligence collection is focused on supporting EW. If not properly coordinated with the G-2 or S-2 staff, electronic attack operations may impact intelligence collection by jamming or inadvertently interfering with a particular frequency being used to collect data on the threat, or by jamming a given enemy frequency or system that deprives friendly forces of that means of collecting data. Either can significantly deter intelligence collection efforts and their ability to answer critical information requirements. Coordination between the EW officer, the fire support coordinator, and the G-2 or S-2 is prevents this interference. In situations where a known conflict between the intelligence collection effort and the use of electronic attack exists, the EW working group brings the problem to the G-3 or S-3 for resolution.

4-65. Other forms of effects rely on electromagnetic spectrum. For example, psychological operations may plan to use a given set of frequencies to broadcast messages, or a military deception plan may include the broadcast of friendly force communications. In both examples, the use of electronic attack could unintentionally interfere or disrupt such broadcasts if not properly coordinated. To ensure electronic attack does not negatively impact planned operations, the EW officer coordinates between fires, network operations, and other functional or integrating cells as required.

4-66. Like any other form of electromagnetic radiation, electronic attack can adversely affect local media and communications systems and infrastructure. EW planners consider unintended consequences of EW operations and deconflict these operations with the various functional or integrating cells. For example, friendly jamming could potentially deny the functioning of essential services such as ambulance or fire fighters to a local population. EW officers routinely synchronize electronic attack with the other functional or integrating cells responsible for the information tasks. In this way, they ensure that electronic attack efforts do not cause fratricide or unacceptable collateral damage to their intended effects.

4-67. The potential for hostile intelligence collection also affects electronic attack. A well-equipped enemy can detect friendly EW capabilities and thus gain intelligence on friendly force intentions. For example, the frequencies Army forces jam could indicate where they believe the enemy's capabilities lie. The EW officer and the G-2 or S-2 develop an understanding of the enemy's collection capability. Along with the red team (if available), they determine what the enemy might gain from friendly force use of electronic attack. (A *red team* is an organizational element comprised of trained and educated members that provide an independent capability to fully explore alternatives in plans and operations in the context of the operational environment and from the perspective of adversaries and others. [JP 2-0])

4-68. The effects of jamming only persist as long as the jammer itself is emitting and is in range to affect the target. Normally this time frame is a matter of seconds or minutes, which makes the timing of such missions critical. This is particularly true when jamming is used in direct support of aviation platforms. For example, in a mission that supports suppression of enemy air defense, the time on target and duration of the jamming must account for the speed of attack of the aviation platform. They must also account for the potential reaction time of enemy air defensive countermeasures. The development of directed-energy weapons may change this dynamic in the future. However, at present (aside from antiradiation missiles), the effects of jamming are less persistent than effects achieved by other means.

ELECTRONIC PROTECTION CONSIDERATIONS

4-69. Electronic protection is achieved through physical security, communications security measures, system technical capabilities (such as frequency hopping and shielding of electronics), spectrum management, and emission control procedures. The EW officer and EW working group members must consider the following key functions when planning for electronic protection operations:

- Vulnerability analysis and assessment.
- Monitoring and feedback.
- Electronic protection measures and how they affect friendly capabilities.

Vulnerability Analysis and Assessment

4-70. Vulnerability analysis and assessment forms the basis for formulating electronic protection plans. The Defense Information Systems Agency operates the Vulnerability Analysis and Assessment Program, which specifically focuses on automated information systems and can be very useful in this effort.

Monitoring and Feedback

4-71. The National Security Agency monitors communications security. Their programs focus on telecommunications systems using wire and electronic communications. Their programs can support and remediate the command's communications security procedures when required.

Electronic Protection Measures and Their Effect on Friendly Capabilities

4-72. Electronic protection measures include any measure taken to protect the force from hostile electronic attack actions. However, these measures can also limit friendly capabilities or operations. For example, denying frequency usage to counter-radio-controlled improvised-explosive-device EW systems on a given frequency to preserve it for a critical friendly information system could leave friendly forces vulnerable to certain radio-controlled improvised explosive devices. The EW officer and the G-6 or S-6 carefully consider these second-order effects when advising the G-3 or S-3 regarding electronic protection measures.

ELECTRONIC WARFARE SUPPORT CONSIDERATIONS

4-73. The distinction between whether a given asset is performing a signals intelligence or EW support mission is determined by who tasks and controls the assets, what they are tasked to provide, and the purpose for which they are tasked. Operational commanders task assets to conduct EW support for the purpose of immediate threat recognition, targeting, planning the conduct of future operations, and other tactical actions (such as threat avoidance and homing). The EW officer coordinates with the G-2 or S-2 to ensure all EW support needed for planned EW operations is identified and submitted to the G-3 or S-3 for approval by the commander. This ensures that the required collection assets are properly tasked to provide the EW support. In cases where planned electronic attack actions may conflict with the G-2 or S-2 intelligence collection efforts, the G-3, S-3, or commander decides which has priority. The EW officer and the G-2 or S-2 develop a structured process within each echelon for conducting this intelligence gain-loss calculus during mission rehearsal exercises and predeployment work-ups.

ELECTRONIC WARFARE REPROGRAMMING CONSIDERATIONS

4-74. Electronic warfare reprogramming refers to modifying friendly EW or target sensing systems in response to validated changes in enemy equipment and tactics or the electromagnetic environment. (See paragraph 1-40 for the complete definition.) Reprogramming EW and target sensing system equipment falls under the responsibility of each Service or organization through its respective EW reprogramming support programs. It includes changes to self-defense systems, offensive weapons systems, and intelligence collection systems. During joint operations, swift identification and reprogramming efforts are critical in a rapidly evolving hostile situation. The key consideration for EW reprogramming is joint coordination. Joint coordination of Service reprogramming efforts ensures reprogramming requirements are identified, processed, and implemented consistently by all friendly forces. During joint operations, EW reprogramming coordination and monitoring is the responsibility of the joint force commander's EW staff. (For more information on EW reprogramming, see FM 3-13.10).

SECTION II — ELECTRONIC WARFARE PREPARATION

4-75. *Preparation* consists of activities performed by units to improve their ability to execute an operation. Preparation includes, but is not limited to, plan refinement; rehearsals; intelligence, surveillance, and reconnaissance; coordination; inspections; and movement (FM 3-0). Preparation creates conditions that improve friendly forces' opportunities for success. It facilitates and sustains transitions, including those to branches and sequels.

4-76. During preparation, the EW officer and members of the EW working group focus their actions on the following activities:

- Revising and refining the EW estimate, EW tasks supporting command and control warfare, and EW support to the overall plan.
- Rehearsing the synchronization of EW support to the plan (including integration into the targeting process, request procedures for joint assets, deconfliction procedures, and asset determination and refinement).
- Synchronizing the collection plan and intelligence synchronization matrix with the attack guidance matrix and EW input to the operation plan or order annexes and appendixes.
- Assessing the planned task organization developed to support EW operations, including liaison officers and organic and nonorganic capabilities required by echelon.
- Coordinating procedures with ISR operational elements (such as signals intelligence staff elements).
- Training the supporting staff members of the EW working group during mission rehearsal exercises.
- Completing precombat checks and inspections of EW assets.
- Completing sustainment preparations for EW assets.
- Coordinate with the G-4 or S-4 to develop EW equipment reporting formats.
- Completing briefbacks by subordinate EW working groups on planned EW operations.
- Refining content and format for the EW officer's portion of the battle update assessment and brief.

SECTION III — ELECTRONIC WARFARE EXECUTION

4-77. *Execution* is putting the plan into action by applying combat power to accomplish the mission and using situational understanding to assess progress and make execution and adjustment decisions (FM 3-0). Commanders focus their subordinates on executing the concept of operations by issuing their intent and mission orders.

4-78. During execution, the EW officer and EW working group members—

- Serve as the EW expert for the commander.
- Maintain the running estimate for EW operations.
- Monitor EW operations and recommend adjustments during execution.
- Recommend adjustments to the commander's critical information requirements based on the situation.
- Recommend adjustments to EW-related control measures and procedures.
- Maintain direct liaison with the fires and network operations cells and the command and control warfare working group (if formed) to ensure integration and deconfliction of EW operations.
- Coordinate and manage EW taskings to subordinate units or assets.
- Coordinate requests for nonorganic EW support.

- Continue to assist the targeting working group in target development and recommend targets for attack by electronic attack assets.
- Receive, process, and coordinate subordinate requests for EW support during operations.
- Receive and process immediate support requests for suppression of enemy air defense or EW from joint or multinational forces; coordinate through fire support officer and fire support coordinator with the battlefield coordination detachment and joint or multinational liaisons for support request.
- Coordinate with airspace control section on all suppression of enemy air defense or EW missions.
- Provide input to the overall assessment regarding effectiveness of electronic attack missions.
- Maintain, update, and distribute the status of EW assets.
- Validate and disseminate cease-jamming requests.
- Coordinate and expedite electromagnetic interference reports with the analysis and control element for targeting and the spectrum manager for potential deconfliction.
- Perform jamming control authority function for ground-based EW within the assigned area of operations (when designated by the jamming control authority).

SECTION IV — ELECTRONIC WARFARE ASSESSMENT

4-79. *Assessment* is the continuous monitoring and evaluation of the current situation, particularly the enemy, and progress of an operation (FM 3-0). Commanders, assisted by their staffs, continuously assess the current situation and progress of the operation and compare it with the concept of operations, mission, and commander's intent. Based on their assessment, commanders direct adjustments, ensuring that the operation remains focused on the mission and commander's intent.

4-80. As depicted in figure 4-5 (page 4-10), assessment occurs throughout every operations process activity and includes three major tasks:

- Continuously assessing the enemy's reactions and vulnerabilities.
- Continuously monitoring the situation and progress of the operation towards the commander's desired end state.
- Evaluating the operation against measures of effectiveness and measures of performance.

4-81. The EW officer and supporting members of the EW working group make assessments throughout the operations process. During planning and preparation activities, assessments of EW are made during the MDMP, IPB, targeting, ISR synchronization, and composite risk management integration.

4-82. The EW officer, in conjunction with the G-5 or S-5, helps develop the measures of performance and measures of effectiveness for evaluating EW operations during execution. A *measure of performance* is a criterion used to assess friendly actions that is tied to measuring task accomplishment (JP 3-0). A *measure of effectiveness* is a criterion used to assess changes in system behavior, capability, or operational environment that is tied to measuring the attainment of an end state, achievement of an objective, or creation of an effect (JP 3-0). In the context of EW, an example of a measure of performance is the percentage of known enemy command and control nodes targeted and attacked by electronic attack means (action) versus the number of enemy command and control nodes that were actually destroyed or rendered inoperable for the desired duration (task accomplishment). Measures of effectiveness are used to determine the degree to which an EW action achieved the desired result. This is normally measured through analysis of data collected by both active and passive means. For example, effectiveness is measured by using radar or visual systems to detect changes in enemy weapons flight and trajectory profiles.

4-83. During execution, the EW officer and members of the EW working group participate in combat assessments within the fires cell to determine the effectiveness of electronic attack employment in support of operations. Combat assessment consists of three elements: munitions effects assessment, battle damage assessment, and reattack recommendations. (Paragraphs 4-47 to 4-49 discuss combat assessment.)

SUMMARY

4-84. The EW officer and staff members supporting the EW working group ensure the successful integration of EW capabilities into operations. The EW officer leads the EW integration effort throughout the operations process. The EW officer must be familiar with and participate in the applicable integrating processes and continuing activities discussed within this chapter.

Chapter 5

Coordination, Deconfliction, and Synchronization

Once the commander approves an operation plan or order and preparations are complete, the electronic warfare officer and supporting staff turn to coordinating, deconflicting, and synchronizing the electronic warfare efforts. They ensure electronic warfare actions are carried out as planned or are modified in response to current operations. This chapter discusses major areas and activities that require continuous coordination, deconfliction, and synchronization by the electronic warfare officer and supporting staff of the electronic warfare working groups.

COORDINATION AND DECONFLICTION

5-1. A certain amount of coordination is part of the planning process. However, once a plan is approved and an operation begins, the electronic warfare (EW) staff effort shifts to the coordination and deconfliction necessary to ensure units carry out EW actions as planned or modify actions to respond to the dynamics of the operation.

5-2. The EW officer and members of the EW working group continuously monitor several key areas. These include EW coordination across organizations (higher, lower, and adjacent units), support request coordination, electromagnetic spectrum management, EW asset management, functional coordination between EW subdivisions, EW reprogramming, and EW deconfliction. Normally, EW personnel on watch in the operations center monitor and coordinate activities of these key areas. They alert the EW officer or other EW support personnel to address the required actions.

COORDINATION ACROSS ORGANIZATIONS

5-3. At the joint level, the information operations division of the J-3 performs EW coordination. The EW section of the information operations staff engages in all EW functions. This section performs peacetime contingency planning, completes day-to-day planning and monitoring of routine theater EW activities, and crisis action planning for contingencies as part of emergent joint operations. The EW section coordinates closely with other appropriate staff sections and other larger joint planning groups as required. (JP 3-13.1 discusses joint EW coordination.)

5-4. In the early stages of contingencies, the joint force commander's EW staff assesses the staffing requirements for planning and execution. This staff also coordinates EW planning and course of action development with the joint force commander's components. Services begin component EW planning and activate their EW working groups per combatant command or Service guidelines. When the scope of a contingency becomes clearer, the command EW officer may request that the joint force commander establish a joint EW coordination cell. If a joint EW coordination cell is formed, it normally requires additional augmentation from the Service or functional components. Depending on the size of the force, EW personnel from the division, corps, or theater are expected to augment the joint EW coordination cell to form a representative EW planning and execution organization. The senior Army organization's staff EW officer anticipates this requirement and prepares to support the augmentation if requested.

5-5. Coordination occurs through established EW working groups from theater level to battalion level. Within Army organizations, the coordination of EW activities occurs both horizontally and vertically. At every level, the staff EW officer ensures the necessary coordination. Normally, coordination of EW activities between the Army and joint force air component commander flows through the battlefield coordination detachment at the joint air operations center. EW staffs at higher echelons monitor EW-related activities and resolve conflicts when necessary.

5-6. Normally the senior Army headquarters (ARFOR) G-3 or S-3 coordinates with external EW organizations, unless direct liaison is authorized at lower echelons. Other components requesting Army EW support coordinate their support requirements with the EW officer located at the ARFOR headquarters or tactical operations center. Often, a liaison from the requesting organization completes these requests. If other Service or functional components have an immediate need for Army EW support, they send the request to the operational fires directorate or fires cell and the senior headquarters EW working group (sometimes referred to as an EW coordination cell) via the Global Command and Control System or Global Command and Control System-Army. In support of external EW coordination, the staff EW officer within the J-3, G-3, or S-3—

- Provides an assessment of EW capabilities to other component operation centers.
- Coordinates preplanned EW operations with other Service components (within prescribed time lines).
- Updates preplanned EW operations in coordination with other components as required.

Support Request Coordination

5-7. Units requesting electronic attack support forward requests to the appropriate EW working group. (See appendix D for the electronic attack request format.) Each EW working group prioritizes the requests and forwards them to the higher headquarters. The commander who owns the capability when the requested support is needed approves the requests. The technical data required to support the execution of the request is passed through EW channels at the appropriate level of classification.

5-8. Electronic warfare support requests are prioritized and passed from the EW working groups through G-2 or S-2 channels and are approved by the commander who owns the capability. New EW support requests are integrated into the intelligence synchronization process. If they are approved, they appear in the intelligence synchronization plan and the unit intelligence, surveillance, and reconnaissance plan. See FMI 2-01 for details on the intelligence synchronization process. The technical data required to support EW support requests passes via signals intelligence channels within the G-2 or S-2 by classified means.

Electromagnetic Spectrum Management

5-9. The electromagnetic spectrum is a finite resource. Once apportioned, this resource must be managed efficiently to maximize the limited spectrum allocated to support military operations. Electromagnetic spectrum operations aim to enable electronic systems to perform their functions in the intended environment without causing or experiencing unacceptable interference. Electromagnetic spectrum operations deconflict all military, national, and host-nation systems being used in the area of operations, including electronic protection systems, communications systems, sensors, and weapon systems.

5-10. Spectrum management involves planning, coordinating, and managing use of the electromagnetic spectrum through operational, engineering, and administrative procedures. Primarily, it involves determining what specific activities will occur in each part of the available spectrum. For example, some frequencies are assigned to the counter radio-controlled improvised-explosive-device EW systems operating in the area of operations. These frequencies then are deconflicted with ground tactical communications. The spectrum manager ensures all necessary functions that require use of the electromagnetic spectrum have sufficient allocation of that spectrum to accomplish their purpose. Where a conflict (two or more functions require the same portion of the spectrum) exists, the spectrum manager resolves the conflict through direct coordination. Figure 5-1 shows the basic procedures the spectrum manager follows to deconflict spectrum use.

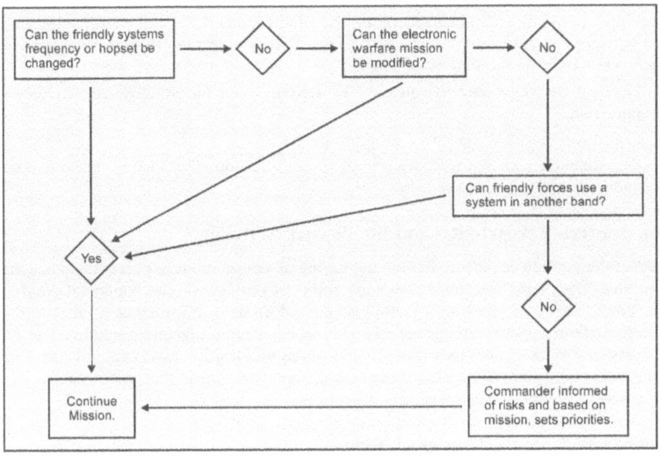

Figure 5-1. Spectrum deconfliction procedures

5-11. The spectrum manager is a member of the G-6 or S-6 section that has staff responsibility for spectrum management in the unit. The spectrum manager is a member of the unit's EW working group. Conflicts regarding spectrum use and allocation that cannot be resolved through direct coordination by the spectrum manager are referred to the G-3 or S-3 for resolution.

Jamming Control Authority

5-12. Depending on the operational situation, an Army headquarters may be designated as the jamming control authority. This authority serves as the senior jamming control authority in the area of operations. It establishes guidance for jamming on behalf of the joint force commander. If designated as the jamming control authority, the senior staff EW officer normally is tasked with the following responsibilities:

- Participating in development of and ensuring compliance with the joint restricted frequency list.
- Validating and approving or denying cease-jamming requests.
- Maintaining situational awareness of all jamming-capable systems in the area of operations.
- Acting as the joint force commander's executive agent for developing EW intelligence gain-or-loss recommendations when electronic attack or electronic warfare support conflicts occur.
- Coordinating jamming requirements with joint force components.
- Investigating unauthorized jamming events and implementing corrective measures.

See JP 3-13.1 for further information on jamming control authority.

Asset Management

5-13. Regardless of echelon, the EW officer monitors and tracks the organization's EW assets and their status. The EW officer makes recommendations to the G-3 or S-3 concerning EW asset allocation and reallocation when required. The EW officer monitors and tracks EW asset status within the EW working group and reports this information to higher echelons via the Army battle command system.

Other Coordinating Actions

5-14. In addition to the functional considerations listed in chapter 4, several coordinating actions must also take place between the EW working groups (at all echelons) and the other planning and execution cells within the headquarters. These actions include—

- Detailed coordination between the EW activities and the intelligence activities supporting an operation.
- Coordination of EW systems reprogramming.
- Coordination with the working groups or cells coordinating the command and control warfare and information protection tasks.

Coordination Between EW Activites and Intelligence Activities

5-15. Most of the intelligence effort, before and during an operation, relies on collection activities targeted against various parts of the electromagnetic spectrum. Electronic warfare support depends on the timely collection, processing, and reporting of intelligence and combat information to alert EW operators and other military activities about intelligence collected in the electromagnetic spectrum. The EW officer and G-2 or S-2 ensure EW collection priorities and EW support collection assets are integrated into a complete intelligence collection plan. This plan ensures that units maximize the use of scarce intelligence and collection assets to support the commander's objectives.

Coordination of EW Systems Reprogramming

5-16. The EW officer and G-2, at division and corps levels, track and coordinate EW systems reprogramming input submitted by lower echelons. This input is then forwarded to the Army Service component command headquarters for submission to the Army Reprogramming Analysis Team. EW officers ensure this input is promptly submitted to ensure urgent reprogramming actions are completed for assigned systems. See FM 3-13.10 for detailed procedures for reprogramming EW and target sensing systems.

Coordination Between EW, Command and Control Warfare, and Information Tasks

5-17. EW working groups coordinate their supporting actions with the elements responsible for the Army information tasks—information engagement, command and control warfare, information protection, operations security, and military deception. Although EW plays a major role in supporting command and control warfare and information protection, it also enhances or provides direct support to other information tasks. For example, enemy radio and television broadcasts can be disrupted or replaced with friendly radio and television messages as part of larger psychological operations in support of information engagement. Electronic deception capabilities can support and enhance an overall military deception operation.

Deconfliction

5-18. Friendly forces depend on electromagnetic energy and the electromagnetic spectrum to sense, process, store, measure, analyze, and communicate information. This dependency creates the potential for significant interference between various friendly systems. Without proper deconfliction, interference could damage friendly capabilities or lead to operational failure. This is especially true with regard to EW systems. EW deconfliction includes—

- Friendly electromagnetic spectrum use for communications and other purposes (such as navigation systems and sensors) with electronic attack activities (such as counter-radio-controlled improvised-explosive-device EW systems).
- Electronic attack activities with electronic warfare support activities (potential electromagnetic interference of collection assets).

- Electronic attack and electronic warfare support activities with information tasks involving electromagnetic emissions (such as counter-radio-controlled improvised-explosive-device EW systems interfering with a psychological-operations radio broadcast).
- Electronic attack activities with host-nation electromagnetic spectrum users (such as commercial broadcasters, emergency first responders, and law enforcement).

5-19. The forum for deconfliction is the unit's EW working group. As such, the specific composition of the working group may expand to include more than the standard staff representation described in chapter 3. Regardless of echelon, to perform its critical deconfliction function, the EW working group retains knowledgeable representation from and ready access to decisionmakers. The EW working group also retains knowledge of and access to higher headquarters assistance and reachback capabilities available (See appendix F for more information).

SYNCHRONIZATION

5-20. EW, particularly in electronic attack, can produce both intended and unintended effects. Therefore, units thoroughly synchronize its use with other forms of fires and with friendly systems operating in the electromagnetic spectrum. Through synchronization, units avoid negative effects such as communications fratricide by jammers. The EW officer ensures all EW activities are integrated into the appropriate sections of plans—fires, information protection, command and control warfare, and military deception plans. This officer also synchronizes EW activities for maximum contribution to the commander's desired effects while preventing EW from inhibiting friendly force capabilities. The primary forum for this synchronization is the unit's EW working group. The EW officer attends the regular targeting meetings in the fires cell and may also participate (perhaps as a standing member) in other functional or integrating cells and working groups. These may include fires, information engagement, network operations, or future operations. The EW officer's participation in these other cells and working groups helps to synchronize EW operations.

SUMMARY

5-21. EW capabilities yield many advantages for the commander. The EW working group's sole purpose is to facilitate the integration, coordination, deconfliction, and synchronization of EW operations to ensure advantages are achieved. This effort requires constant coordination with the unit's other functional cells and working groups. As conflicts are identified during the planning and execution of operations, the EW officer and supporting staff members coordinate solutions to those conflicts within the EW working group.

Chapter 6

Integration with Joint and Multinational Operations

Joint warfare is team warfare. It requires the integrated and synchronized application of all appropriate capabilities. During joint operations, Services work together to accomplish a mission. In multinational operations, forces of two or more nations work together to accomplish a mission. During both joint and multinational operations, forces operate under established organizational frameworks and coordination guidelines. This chapter describes the joint and multinational operational frameworks and guidelines for integrating electronic warfare capabilities.

JOINT ELECTRONIC WARFARE OPERATIONS

6-1. One strength of operating as a joint force is the ability to maximize combat capabilities through unified action. However, the ability to maximize the capabilities of a joint force requires guidelines and an organizational framework that can be used to integrate them effectively. JP 3-13.1 establishes the guidelines and organizational framework for joint electronic warfare (EW) operations.

6-2. Joint task forces are task-organized. Therefore, their composition varies based on the mission. Normally the EW organization within a joint force centers on the—

- Component commands.
- Supporting joint centers.
- Joint force staff.
- Joint force commander's EW staff, joint electronic warfare coordination cell, or information operations (IO) cell.

The supporting centers for EW operations may include the joint operations center, joint intelligence center, Joint Frequency Management Office (JFMO), and joint targeting coordination board.

JOINT FORCE PRINCIPAL STAFF FOR ELECTRONIC WARFARE

6-3. In EW, the principal staff consists of the J-2, J-3, and J-6. The J-2 collects, processes, tailors, and disseminates all-source intelligence for EW. The J-3 has primary staff responsibility for EW activity. This director also plans, coordinates, and integrates joint EW operations with other combat disciplines in the joint task force. Normally, the joint force commander's EW staff or a joint EW coordination cell and an IO cell assist the J-3. The joint force staff network operations director (in the J-6) coordinates electromagnetic spectrum use for information systems with electromagnetic-dependent weapons systems used by the joint force. The IO officer is the principal IO advisor to the J-3. This officer is the lead planner for integrating, coordinating, and executing IO. The command EW officer is the principal EW planner on the J-3 staff. This officer coordinates with the IO cell to integrate EW operations fully with other IO core, supporting, and related capabilities (see JP 3-13.1 for further information)

JOINT FORCE COMMANDER'S ELECTRONIC WARFARE STAFF

6-4. A joint force commander's EW staff supports the joint force commander in planning, coordinating, synchronizing, and integrating joint force EW operations. The joint force commander's EW staff ensures that joint EW capabilities support the joint force commander's objectives. The joint force commander's EW staff is an element within the J-3. It consists of representatives from each component of the joint force.

An EW officer appointed by the J-3 leads this element. The joint force commander's EW staff includes representatives from the J-2 and J-6 to facilitate intelligence support and EW frequency deconfliction.

6-5. On many joint staffs, the intra-staff coordination previously accomplished through a joint force commander's EW staff is now performed by an IO cell or similar organization. An IO cell, if established, coordinates EW activities with other IO activities to maximize effectiveness and prevent mutual interference. If both a joint force commander's EW staff and an IO cell exist, a joint force commander's EW staff representative may be assigned to the IO cell to facilitate coordination. For more information about the organization and procedures of the joint IO cell, see JP 3-13.

JOINT ELECTRONIC WARFARE COORDINATION CELL

6-6. The decision to form a joint EW coordination cell depends on the anticipated role of EW in an operation. When EW is expected to play a significant role in the joint force commander's mission, a component command's EW coordination organization may be designated as the joint EW coordination cell to handle the EW aspects of the operation. The joint EW coordination cell may be part of the joint force commander's staff, be assigned to the J-3 directorate, or remain within the designated component commander's structure. The joint EW coordination cell plans operational-level EW for the joint force commander. (JP 3-13.1 discusses the joint EW coordination cell in more detail.)

JOINT TASK FORCE COMPONENT COMMANDS

6-7. Joint task force component commanders exercise operational control of their EW assets. Each component is organized and equipped to perform EW tasks in support of its basic mission and to provide support to the joint force commander's overall objectives. If a component command (Service or functional) is designated to stand up a joint EW coordination cell, it executes the responsibilities and functions outlined in JP 3-13.1.

6-8. A major consideration for standing up a joint EW coordination cell at the component command level is access to a special compartmented information facility to accomplish the cell's required coordination functions. Optimal joint EW coordination cell staffing dictates including special technical operations personnel cleared to coordinate and deconflict special technical operations issues. Special technical operations are associated with the planning and coordination of advanced special programs and the integration of new capabilities into operational units.

6-9. Under current force structure, the special technical operations requirement limits the activation of a joint EW coordination cell to organizations at corps and above levels. Organizations below corps level require significant joint augmentation to meet the special technical operations requirement.

JOINT FREQUENCY MANAGEMENT OFFICE

6-10. Joint policy tasks each geographic combatant commander to establish a structure to manage spectrum use and establish procedures that support ongoing operations. This structure must include a JFMO. The JFMO may be assigned from the supported combatant commander's J-6 staff, from a component's staff, or from an external command such as the Joint Spectrum Center. The JFMO coordinates the information systems use of the electromagnetic spectrum, frequency management, and frequency deconfliction. The JFMO develops the frequency management plan and makes recommendations to alleviate mutual interference.

6-11. The G-6 or S-6 coordinates the Army's use of the electromagnetic spectrum, frequency management, and frequency deconfliction with the JFMO through the network operations cell. If established, coordination with the joint spectrum management element is required. (See figure 6-1.)

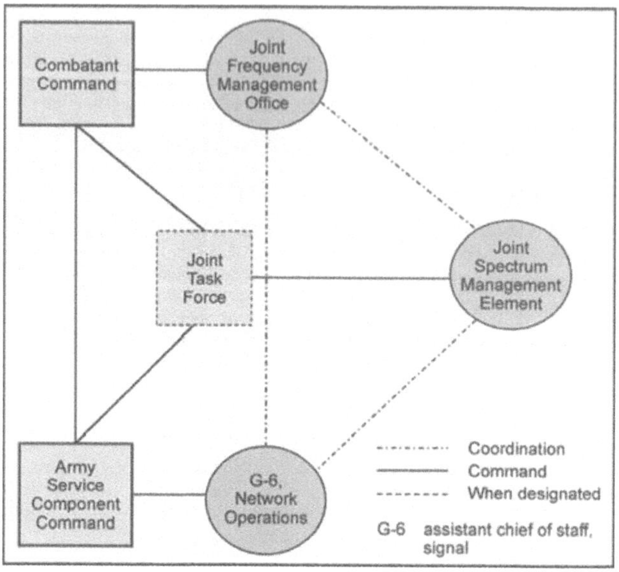

Figure 6-1. Joint frequency management coordination

JOINT INTELLIGENCE CENTER

6-12. The joint intelligence center is the focal point for the intelligence structure supporting the J-2. Directed by the J-2, the joint intelligence center communicates directly with component intelligence agencies and monitors intelligence support to EW operations. This center can adjust intelligence gathering to support EW missions. Within the G-2, EW support requests are coordinated through the requirement cell and then forwarded to the requirements division within the joint intelligence center. (See figure 6-2, page 6-4.)

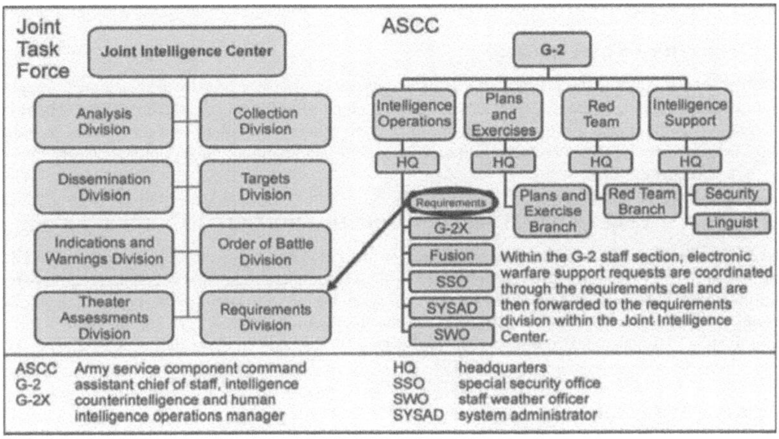

Figure 6-2. Electronic warfare support request coordination

6-13. The composition and focus of each joint intelligence center varies by theater. However, each can perform indications and warning as well as collect, manage, and disseminate current intelligence. Through the joint intelligence center, the ARFOR (Army Service component) headquarters coordinates support from the Air Force, Navy, and Marine Corps and national, interagency, and multinational sources. In addition to its other functions, the joint intelligence center coordinates the acquisition of national intelligence for the joint task force and the combatant command's staff.

JOINT TARGETING COORDINATION BOARD

6-14. The joint targeting coordination board focuses on developing broad targeting priorities and other targeting guidance in accordance with the joint force commander's objectives as they relate operationally. The joint targeting coordination board remains flexible enough to address targeting issues without becoming overly involved in tactical-level decisionmaking. Briefings conducted at the joint targeting coordination board focus on ensuring that intelligence, operations (by all components and applicable staff elements), fires, and maneuver are on track, coordinated, and synchronized. For further information on the joint targeting coordination board, see JP 3-60.

MULTINATIONAL ELECTRONIC WARFARE OPERATIONS

6-15. EW is an integral part of multinational operations (sometimes referred to as combined operations). U.S. planners integrate U.S. and multinational EW capabilities into a single, integrated EW plan. U.S. planners provide multinational forces with information concerning U.S. EW capabilities and provide them EW planning and operational support. However, the planning of multinational force EW is difficult due to security issues, differences in levels of training, language barriers, and terminology and procedural issues. U.S. and North Atlantic Treaty Organization (NATO) EW doctrine provide commonality and a framework for using EW in NATO operations. (See Allied Joint Publication 3.6 for specific information.)

MULTINATIONAL FORCE COMMANDER

6-16. The multinational force commander provides guidance for planning and conducting EW operations to the multinational force through the C-3 and the EW coordination cell. The EW coordination cell is located at multinational force headquarters. An IO cell may also be established to coordinate all IO-related activities, including related EW operations.

Joint Operations Staff Section

6-17. Within the multinational staff, the joint operations section has primary responsibility for planning and integrating EW activities. A staff EW officer is designated with specific responsibilities. These include integrating multinational augmentees, interpreting or translating EW plans and procedures, coordinating appropriate communications connectivity, and integrating multinational force communications into a joint restricted frequency list.

Multinational Electronic Warfare Coordination Cell

6-18. In multinational operations, the multinational force commander uses an EW coordination cell as the mechanism for coordinating EW resources within the area of operations. This cell is an integral part of the multinational joint force headquarters J-3 staff, at whatever level is appropriate. It provides an effective means of coordinating all EW activities by the multinational force. The multinational force EW coordination cell plans and coordinates all in-theater EW activities in close liaison with the J-2, J-5, and J-6.

Electronic Warfare Mutual Support

6-19. Electronic warfare mutual support is the timely exchange of EW information to make the best use of the available resources. It is facilitated by the use of an agreed reference database called the NATO emitter database. Electronic warfare mutual support procedures developed during EW planning include—

- A review of friendly and enemy information data elements that may be exchanged.
- Mechanisms leading to the exchange of data during peace, crisis, and war.
- Development of peacetime exercises to practice the exchange of data.
- Establishment of EW points of contact with adjacent formations and higher and subordinate headquarters for planning purposes, regardless of whether EW resources exist or not.
- Initial acquisition and maintenance of multinational force EW capabilities.
- Exchange of EW liaison teams equipped with appropriate communications.
- Establishment and rehearsal of contingency plans for the exchange of information on friendly and enemy forces.
- Development of communications protocols in accordance with NATO Standardization Agreement (STANAG) 5048.
- Provision of secure, dedicated, and survivable communications.

Other Considerations

6-20. EW in multinational operations addresses other considerations. Soldiers must consider—

- Exchange of EW information.
- Exchange of signals intelligence information.
- Exchange of the electronic order of battle.
- Electronic warfare reprogramming.

6-21. Army forces participating in multinational EW operations must exchange EW information with other forces. They must help develop joint information exchange protocols and use those protocols for conducting operations.

6-22. Exchanging signals intelligence information requires care to avoid violating signals intelligence security rules. The policy and relationship between EW and signals intelligence within NATO are set out in NATO Military Committee (MC) 64.

6-23. In peacetime, before forming a multinational force, the exchange of electronic order of battle information is normally achieved under bilateral agreement. During multinational operations, a representative of the joint EW coordination cell, through the theater joint analysis center or the joint intelligence center, ensures the maintenance of an up-to-date electronic order of battle. The inclusion of multinational forces is based on security and information exchange guidelines agreed upon by the participating nations.

6-24. Electronic warfare reprogramming is a national responsibility. However, the joint EW coordination cell remains aware of reprogramming efforts being conducted within the multinational force. FM 3-13.10 guides the Army's reprogramming effort.

SUMMARY

6-25. Every joint or multinational operation is uniquely organized to accomplish the mission. Army EW officers integrate EW forces and capabilities with the organizations and agencies outlined in this chapter. To coordinate Army EW operations with joint and multinational forces, Army EW officers must understand fully the organizational frameworks, policies, and guidelines established for joint and multinational EW operations.

Chapter 7
Electronic Warfare Capabilities

Electronic warfare capabilities consist of high-demand, low-density assets across the Services. Hence, the conduct of electronic warfare operations requires joint interdependence. This complex interdependence extends beyond the traditional Service capabilities. It includes national agencies—such as the Central Intelligence Agency, National Security Agency, and Defense Intelligence Agency—that constantly seek to identify, catalog, and update the electronic order of battle of enemies and adversaries. To support the joint force commander, the subject matter expertise and unique capabilities provided by each Service, agency, and branch or proponent are integrated with all available electronic warfare capabilities.

SERVICE ELECTRONIC WARFARE CAPABILITIES

7-1. Each Service maintains electronic warfare (EW) capabilities to support operational requirements. During operations, the Army is dependent on organic and nonorganic EW capabilities from higher echelons, joint forces, and national agencies. Army EW planners leverage all available EW capabilities to support Army operations. Although not all-inclusive, appendix E provides a listing of current Army, Marine Corps, Navy, and Air Force EW capabilities and references.

EXTERNAL SUPPORT AGENCIES AND ACTIVITIES

7-2. Army EW planners routinely use and receive support from external organizations to assist in planning and integrating EW operations. Support from these organizations may include personnel augmentation, functional area expertise, technical support, and planning support.

Big Crow Program Office

7-3. The Big Crow Program Office was established in 1971 to provide testing environments for U.S. military radio frequency sensor, communication, and navigation systems. Today, the Big Crow Program Office provides customers with joint, multifunctional support for testing communications, sensors, information operations, and related weapon systems in support of Department of Defense (DOD), the individual Services, the National Aeronautics and Space Administration, the National Reconnaissance Office, and others. This support includes replicating information operations and EW threat environments as well as providing telemetry recording, technology prototyping, proof-of-concept demonstrations, and information operations and EW training. Big Crow's mission and capabilities now span the electromagnetic spectrum, encompassing EW, telemetry, radar, and electro-optical systems. Mobile and worldwide deployable, the Big Crow Program Office offers a variety of capabilities.

Defense Information Systems Agency

7-4. The Defense Information Systems Agency is a combat support agency. It plans, develops, fields, operates, and supports command, control, communications, and information systems. These systems serve the President, the Secretary of Defense, the Joint Chiefs of Staff, the combatant commanders, and other DOD components. The Defense Information Systems Agency also operates the Vulnerability Analysis and Assessment Program. This program specifically focuses on automated information systems.

Joint Communications Security Monitor Activity

7-5. The Joint Communications Security Monitor Activity was created in 1993 by a memorandum of agreement between the Services' operations deputies, Directors of the Joint Staff, and the National Security Agency. The Joint Communications Security Monitor Activity monitors (collects, analyzes, and reports) communications security of DOD telecommunications and automated information systems as well as related noncommunications signals. Its purpose is to identify potentially exploitable vulnerabilities and to recommend countermeasures and corrective actions. The Joint Communications Security Monitor Activity supports real world operations, joint exercises, and DOD systems monitoring.

Joint Information Operations Warfare Command

7-6. The Joint Information Operations Warfare Command (JIOWC) was activated in 2006 as a functional component to the United States Strategic Command (USSTRATCOM). JIOWC integrates joint information operations into military plans, exercises, and operations across the spectrum of conflict. It is a valuable resource for commanders during the planning and execution of joint information operations. JIOWC deploys information operations planning teams when the commander of USSTRATCOM approves a request for support. This center delivers tailored, highly skilled support and sophisticated models and simulations to joint commanders and provides information operations expertise in joint exercises and contingency operations.

7-7. JIOWC also fields the Joint Electronic Warfare Center. This center provides specialized expertise in EW. It is an innovation center for existing and emerging EW capabilities and tactics, techniques, and procedures via a network of units, labs, test ranges, and academia. The Joint Electronic Warfare Center also has EW reprogramming oversight responsibilities for the Joint Staff. This oversight includes organizing, managing, and exercising joint aspects of EW reprogramming and facilitating the exchange of joint EW reprogramming data. The actual reprogramming of equipment, however, is a Service responsibility.

Joint Spectrum Center

7-8. The Joint Spectrum Center was activated in 1994 under the direction of the joint staff's J-6. The Joint Spectrum Center assumed all the missions and responsibilities previously performed by the Electromagnetic Compatibility Center plus additional responsibilities. Personnel in the Joint Spectrum Center are experts in spectrum planning, electromagnetic compatibility and vulnerability, electromagnetic environmental effects, information systems, modeling and simulation, operations support, and system acquisition. The Joint Spectrum Center provides complete, spectrum-related services to combatant commanders, Services, and other government agencies. The Joint Spectrum Center deploys teams in support of the combatant commanders and serves as the DOD focal point for supporting spectrum supremacy aspects of information operations. It assists Soldiers in developing and managing the joint restricted frequency list and helps to resolve operational interference and jamming incidents. The Joint Spectrum Center can also provide databases of friendly force command and control systems for use in planning electronic protection. The Joint Spectrum Center is a field office within the Defense Spectrum Organization under the Defense Information Systems Agency.

Joint Warfare Analysis Center

7-9. The Joint Warfare Analysis Center is a Navy-sponsored joint command under the J-3 established in 1994. The Joint Warfare Analysis Center assists the Chairman of the Joint Chiefs of Staff and combatant commanders in preparing and analyzing joint operational plans. It provides analysis of engineering and scientific data and integrates operational analysis with intelligence.

Marine Corps Information Technology and Network Operations Center

7-10. The Marine Corps Information Technology and Network Operations Center is the Marine Corps' enterprise network operations center. The Marine Corps Information Technology and Network Operations Center is the nerve center for the central operational direction and configuration management of the Marine

Corps enterprise network. It is co-located with the Marine Corps forces computer network defense, the component to the joint task force for computer network operations, and the Marine Corps computer incident response team. This relationship provides a strong framework for integrated network management and defense.

NATIONAL SECURITY AGENCY

7-11. The National Security Agency/Central Security Service is America's cryptologic organization. This organization protects U.S. government information systems and produces foreign signals intelligence information. Executive Order 12333, 4 December 1981, describes the responsibility of the National Security Agency/Central Security Service in more detail. The resources of National Security Agency/Central Security Service are organized for two national missions:

- The Information Assurance Mission provides the solutions, products, and services, and conducts defensive information operations, to achieve information assurance for information infrastructures critical to U.S. national security interests.
- The Signals Intelligence Mission allows for an effective, unified organization and control of all the foreign signals collection and processing activities of the United States. The National Security Agency is authorized to produce signals intelligence in accordance with objectives, requirements, and priorities established by the Director of National Intelligence in consultation with the President's Foreign Intelligence Advisory Board.

7-12. The Director, National Security Agency is the principal signals intelligence and information security advisor to the Secretary of Defense, Director of National Intelligence, and the Chairman of the Joint Chiefs of Staff. The Director, National Security Agency provides signals intelligence support to combatant commanders and others in accordance with their expressed formal requirements.

SUMMARY

7-13. This chapter and appendix E provide a sampling of available joint and Service EW capabilities, activities, and agencies that support ground force commanders in full spectrum operations. To leverage these capabilities for EW support, Army EW officers acquire a working knowledge of the capabilities available and the procedures for requesting support. Additionally, appendix F provides information on available EW related tools and other resources.

Appendix A
The Electromagnetic Environment

Electromagnetic energy is both a natural and manmade occurrence. This energy, in the form of electromagnetic radiation, consists of oscillating electric and magnetic fields and is propagated at the speed of light. Electromagnetic radiation is measured by the frequency of its wave pattern's repetition within a set unit of time. The standard term for the measurement of electromagnetic radiation is the hertz (Hz), the number of repetitions (cycles) per second. The electromagnetic spectrum refers to the range of frequencies of electromagnetic radiation.

OVERVIEW OF THE ELECTROMAGNETIC ENVIRONMENT

A-1. The electromagnetic environment is the resulting product of the power and time distribution, in various frequency ranges, of radiated or conducted electromagnetic emission levels. Within their intended operational environment, a military force, system, or platform may encounter these emissions while performing tasks during operations. The electromagnetic environment is the sum of—

- Electromagnetic interference.
- Electromagnetic pulse.
- Hazards of electromagnetic radiation to personnel, ordnance, and volatile materials.
- Natural phenomena effects of lightning and precipitation static. (*Precipitation static* is charged precipitation particles that strike antennas and gradually charge the antenna, which ultimately discharges across the insulator, causing a burst of static [JP 3-13.1]).

THE ELECTROMAGNETIC SPECTRUM

A-2. The *electromagnetic spectrum* is the range of frequencies of electromagnetic radiation from zero to infinity. It is divided into 26 alphabetically designated bands (JP 1-02). The spectrum is a continuum of all electromagnetic waves arranged according to frequency and wavelength. The electromagnetic spectrum extends from below the frequencies used for modern radio (at the long-wavelength end) through gamma radiation (at the short-wavelength end). It covers wavelengths from thousands of kilometers to a fraction of the size of an atom. Figure A-1 shows the spectrum regions and wavelength segments associated with the electromagnetic spectrum.

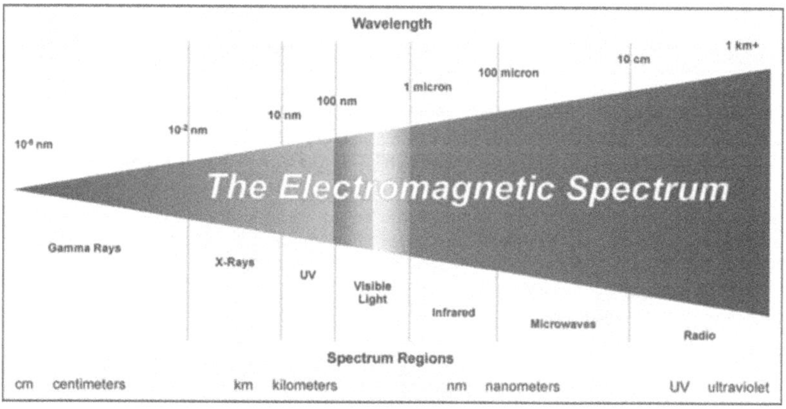

Figure A-1. The electromagnetic spectrum

A-3. Included within the radio and microwave regions of the electromagnetic spectrum are the radio frequency and radar bands. These bands are routinely referred to by their band designators. For example, high frequency radios are HF radios and K-band radars are radars that operate between 18 and 27 gigahertz. Civilian agencies and military forces throughout the world use several different designator systems, which can result in confusion. Table A-1 shows the radio frequency band designators and their associated frequency ranges. It also shows radar band designators, associated frequency ranges, and typical usage. These are standard designations used by the United States.

Table A-1. Radio and radar designators and frequency bands

Radio Frequency Band Designator	Radio Frequency Range	Radar Band Designator*	Frequency Range	Typical Usage
ULF	lower than 3 Hz	VHF	50-330 MHZ	Very long-range surveillance
ELF	3 Hz - 3 kHz	UHF	300-1,000 MHz	Very long-range surveillance
VLF	3 - 30 kHz	L	1-2 Ghz	Long-range surveillance, enroute traffic control
LF	30 - 300 kHz	S	2-4 Ghz	Moderate-range surveillance, terminal traffic control, long-range weather
MF	300 kHz - 3 MHz	C	4-8 Ghz	Long-range tracking, airborne weather
HF	3 - 30 MHz	X	8-12 Ghz	Short-range tracking, missile guidance, mapping, marine radar, airborne intercept
VHF	30 - 300 MHz	K_u	12-18 Ghz	High resolution mapping, satellite altimetry
UHF	300 MHZ - 3 GHz	K	18-27 Ghz	Little use
SHF	3 - 30 GHz	K_a	27-40 Ghz	Very high resolution mapping, airport surveillance
EHF	30 - 300 GHz			
Sub-millimeter	300 Ghz - 1 THz			

EHF extremely high frequency
ELF extremely low frequency
GHz Gigahertz
HF high frequency
Hz hertz
kHz kilohertz
LF low frequency
MF medium frequency
MHz megahertz
SHF super high frequency
THz terahertz
UHF ultra high frequency
ULF ultra low frequency
VHF very high frequency

* Radar band designators relate back to the early development of radar in World War II when the letter designators were used for purposes of secrecy. After the requirement for secrecy was no longer needed, these letter band designators remained.

MILITARY OPERATIONS AND THE ELECTROMAGNETIC ENVIRONMENT

A-4. The impact of the electromagnetic environment upon the operational capability of military forces, equipment, systems, and platforms is referred to as electromagnetic environmental effects. Electromagnetic environmental effects encompass all electromagnetic disciplines, including—

- Electromagnetic compatibility and electromagnetic interference.
- Electromagnetic vulnerability.
- Electromagnetic pulse.
- Electronic protection.
- Hazards of electromagnetic radiation to personnel, ordnance, and volatile materials (such as fuels).
- Natural phenomena effects of lightning and precipitation static.

A-5. *Electromagnetic vulnerability* consists of the characteristics of a system that cause it to suffer a definite degradation (incapability to perform the designated mission) as a result of having been subjected to a certain level of electromagnetic environmental effects (JP 3-13.1). Electronic warfare support plays a key role in identifying the electromagnetic vulnerability of an adversary's electronic equipment and systems. Friendly forces take advantage of these vulnerabilities through electronic warfare operations.

DIRECTED ENERGY

A-6. Directed energy refers to technologies that produce of a beam of concentrated electromagnetic energy or atomic or subatomic particles (see chapter 1). *Directed-energy warfare* is military action involving the use of directed-energy weapons, devices, and countermeasures to either cause direct damage or destruction of enemy equipment, facilities, and personnel, or to determine, exploit, reduce, or prevent hostile use of the electromagnetic spectrum through damage, destruction, and disruption. It also includes actions taken to protect friendly equipment, facilities, and personnel and retain friendly use of the electromagnetic spectrum (JP 1-02). A *directed-energy weapon* is a system using directed energy primarily as a direct means to damage or destroy enemy equipment, facilities, and personnel (JP 1-02). In addition to destructive effects, directed-energy weapons can also support area denial, crowd control, and obscuration.

A-7. The application of directed energy includes lasers, radio-frequency weapons, and particle-beam weapons. As directed-energy weapons evolve, the tactics, techniques, and procedures for their use also evolve to ensure their safe, effective employment. In electronic warfare, most directed-energy applications fit into the category of electronic attack. However, other applications can be categorized as electronic protection or even electronic warfare support. Examples include the following:

- Applications used for electronic attack, which may include—
  - A laser designed to blind or disrupt optical sensors.
  - A millimeter wave directed-energy weapon used for crowd control.
  - A laser-warning receiver designed to initiate a laser countermeasure to defeat a laser weapon.
  - A millimeter wave obscuration system used to disrupt or defeat a millimeter wave system.
  - A device used to counter radio-controlled improvised explosive devices.
- A laser-warning receiver designed solely to detect and analyze a laser signal is used for electronic warfare support.
- A visor or goggle designed to filter out the harmful wavelength of laser light is used for electronic protection.

A-8. As the use of destructive directed-energy weapons grows, Army forces require the capability to collect information on them. Additionally, Army forces require tactics, techniques, and procedures to mitigate directed-energy weapon effects. Currently, the definitions and terms relating to directed energy are articulated within electronic warfare doctrine. As the technologies related to directed energy expand, joint and Army doctrine may discuss employing directed energy under other doctrinal subjects.

Appendix B

Electronic Warfare Input to Operation Plans and Orders

This appendix discusses electronic warfare input to Army and joint plans and orders.

ARMY PLANS AND ORDERS

B-1. This paragraph lists the electronic warfare (EW) information required for Army operation plans and orders. (See figure B-1 on page B-2 for the EW appendix format.) This discussion is based on current doctrine from FM 5-0. When it is republished, FM 5-0 will state where to place EW-related information in the revised plans and orders format. In addition to the appendix 4 (Electronic Warfare) to Annex P (Information Operations), the following components of operation plans and orders may require EW input:

- **Base order or plan**:
  - Sub-subparagraph (2) (Fires) to subparagraph a (Concept of Operations) to paragraph 3 (Execution).
  - Sub-subparagraph (7) (Information Operations) to subparagraph a (Concept of Operations) to paragraph 3 (Execution).
- **Annex D (Fire Support)**:
  - Sub-subparagraph (4) (Electronic Warfare) to subparagraph b (Air Support) to paragraph 3 (Execution)
  - Appendix 1 (Air Support).
- **Annex L (Intelligence, Surveillance, and Reconnaissance)**:
  - Sub-subparagraph (2) (Fires) to subparagraph a (Concept of Operations) to paragraph 3 (Execution).
  - Sub-subparagraph (7) (Information Operations) to subparagraph a (Concept of Operations) to paragraph 3 (Execution).
- **Annex N (Space)**: Sub-subparagraph (10) (Electronic Warfare) to subparagraph b (Space Activities) to paragraph 3 (Execution).
- **Annex P (Information Operations)**:
  - Sub-sub-subparagraph (d) (Electronic Warfare) to sub-subparagraph (8) to subparagraph a (Concept of Support) to paragraph 3 (Execution).
  - Sub-subparagraph (3) (List of Tasks to Electronic Warfare Units) to subparagraph b (Tasks to Subordinate Units) to paragraph 3 (Execution).

[Classification]

Appendix 4 (Electronic Warfare) to Annex P (Information Operations) to OPORD No____

1. SITUATION.

   a. Enemy.
   - Identify the vulnerabilities of enemy information systems and electronic warfare systems.
   - Identify the enemy capability to interfere with accomplishment of the electronic warfare mission.

   b. Friendly.
   - Identify friendly electronic warfare assets and resources that affect electronic warfare planning by subordinate commanders.
   - Identify friendly foreign forces with which subordinate commanders may operate.
   - Identify potential conflicts within the friendly electromagnetic spectrum, especially if conducting joint or multinational operations. Identify and de-conflict methods and priority of spectrum distribution.

   c. Attachments and detachments.
   - List the electronic warfare assets that are attached or detached.
   - List the electronic warfare resources available from higher headquarters.

2. MISSION. State how electronic warfare will support the commander's objectives.

3. EXECUTION.

   a. Scheme of support. State the electronic warfare tasks.

   b. Tasks to subordinate units. Identify the electronic warfare tasks for each unit.

   c. Coordinating instructions.
   - Identify electronic warfare instructions applicable to two or more units.
   - Identify the requirements for the coordination of electronic warfare actions between units.
   - Identify the emission control guidance.

4. SERVICE SUPPORT. Identify service support for electronic warfare operations.

5. COMMAND AND SIGNAL.

   a. Command.

   b. Signal. Identify if any, the special or unusual electronic warfare-related communications requirements.

[Classification]

Figure B-1. Appendix 4 (Electronic Warfare) to annex P (Information Operations) instructions

JOINT PLANS AND ORDERS

B-2. If required to provide EW input to portions of a joint order, the primary areas for input are the following:

- Paragraph 3 (Execution) to appendix 3 (Information Operations) to Annex C (Operations).
- Tab B (Electronic Warfare) to appendix 3 (Information Operations) to Annex C (Operations).

B-3. See CJCSM 3122.03C for the Joint Operations Planning and Execution System format.

Appendix C

Electronic Warfare Running Estimate

This appendix discusses the electronic warfare running estimate. A *running estimate* is a staff section's continuous assessment of current and future operations to determine if the current operation is proceeding according to the commander's intent and if future operations are supportable (FM 3-0).

C-1. The electronic warfare (EW) running estimate is used to support the military decisionmaking process during planning and execution. During planning, the EW running estimate provides an assessment of the supportability of each proposed course of action from an EW perspective. The format of the EW running estimate closely parallels the steps of the military decisionmaking process. It serves as the primary tool for recording the EW officer's assessments, analyses, and recommendations for EW operations. The EW officer and staff in the EW working group are responsible for conducting the analysis and providing recommendations based on the EW running estimate.

C-2. A complete EW running estimate should contain the information necessary to answer any question the commander may pose. If there are gaps in the EW running estimate, the staff identifies the gaps as information requirements and submits them to the intelligence cell. The EW running estimate can form the basis for EW input required in other applicable appendixes and annexes within operation plans and orders. Figure C-1 on page C-2 provides a sample EW running estimate for use during planning.

1. **MISSION.** Show the restated mission resulting from mission analysis.

2. **SITUATION AND CONSIDERATIONS.**

   a. Characteristics of the area of operations.
      - Weather. State how the weather may impact EW operations.
      - Terrain. State how aspects of the terrain may impact EW operations.
      - Civil Considerations. State how rules of engagement and civil emergency responder frequency restrictions may impact EW operations.

   b. Enemy forces. Discuss enemy dispositions, composition, strength, capabilities, and courses of action (COAs) as they affect EW operations. Identify enemy EW vulnerabilities.

   c. Friendly forces.
      - List the current status of the forces EW resources.
      - List the current status of additional EW support resources.
      - Provide a comparison of EW support requirements with available capabilities and recommend solutions for any discrepancies.
      - Identify friendly forces EW vulnerability and recommend solutions.

   d. Assumptions. List any assumptions used that may affect the employment of EW capabilities.

3. **COURSES OF ACTION.**

   a. List the friendly COAs that were wargamed.
   b. List the evaluation criteria identified during the COA analysis.

4. **ANALYSIS.** Analyze each COA using the evaluation criteria identified during COA analysis.

5. **COMPARISION.** Compare each COA. Rank order the COAs for each EW key consideration identified.

6. **RECOMMENDATION AND CONCLUSIONS.** This paragraph translates the "best" course of action (as determined in paragraph 5) into a complete recommendation. It should outline who, what, where, when, how, and why from the EW point of view. It states which course of action can best be supported by friendly EW, and is less vulnerable to enemy EW force capabilities.

   a. Recommend the most supportable COA from an EW perspective.
   b. List any EW related issues, deficiencies and risks and provide recommendations to reduce their impact.

ANNEXES: Include annexes as required. Annexes with pertinent details should be used to the extent practical to support the contents of the estimate. These annexes may be in considerable detail with only the high points included in the body of the estimate. Annexes should add depth to the contents of the estimate, but should not be used as a substitute for key points that should be included in the body of the estimate.

Figure C-1. Example of an electronic warfare running estimate

C-3. Once the commander approves the order, the EW running estimate is used to inform current and future operations. During execution the EW running estimate is used to help determine if current EW operations are proceeding according to plan and if future EW operations are supportable. Figure C-2, page C-3, shows a sample of the information that might be used to update the EW running estimate during execution. The EW officer and supporting staff members within the EW working group produce and update the running estimate.

Current operation order and fragmentary orders

- Define the battlefield environment. Focus on the aspects of the terrain and weather that could assist or enable electronic warfare operations from both a friendly and threat viewpoint.
  - Maintain updated weather and terrain data.
  - Locate terrain for communications and non-communications sites, line of sight.
  - Identify aspects of terrain and weather that may have an impact on the electromagnetic spectrum.
- Define the threat.
  - Communications systems, including threat radio nets and network nodes.
  - Noncommunications emitters.
  - Electronic support systems.
  - Electronic attack systems.
- Identify host-nation use of the electromagnetic spectrum (restricted frequencies such as government, industry, and emergency responders).
- Identify friendly capabilities, shortfalls and readiness.
  - Electronic attack capabilities and status (joint and Army).
  - Location and availability of organic friendly electronic warfare capabilities (such as Prophets and counter-radio-controlled IED EW systems.
  - Electronic warfare vulnerabilities.
  - Equipment updates, both hardware and software.
- Identify enemy capabilities, shortfalls and readiness.
  - Electronic warfare capabilities and status (if known).
  - Electronic warfare vulnerabilities.
  - Electronic order of battle.

Electronic Warfare Target Folder

- Describe the targeted capability, its associated vulnerabilities, and the friendly capabilities used to engage them.
- Maintain updated high-value target and high-payoff target lists.
- Develop a prioritized target list based on high-value targets and high-payoff targets.
- EW target folders are split between traditional and asymmetric targets.
  - Traditional targets might include integrated air defense systems, communications nodes, and radar facilities. Traditional targets are normally fixed or less mobile than asymmetric targets and are easier to develop.
  - Asymmetric targets might include individual cell phones, radio-controlled improvised explosive devices, global positioning systems and wireless networks. Asymmetric targets can be either stationary or mobile and are typically harder to develop than traditional targets during the targeting phase.

Figure C-2. Sample update information to the electronic warfare running estimate

Appendix D

Electronic Warfare-Related Reports and Messages

This appendix provides information and references for electronic warfare and electronic warfare-related reports and message formats.

MESSAGES AND SUMMARIES

D-1. The following messages and summaries are associated with the planning, synchronization, deconfliction, and assessment of EW operations.

ELECTRONIC ATTACK DATA MESSAGE

D-2. An electronic attack data message reports an electronic attack strobe from an affected or detecting unit's position to an aircraft emitting an electronic attack. It is used to determine the location of a hostile or unknown aircraft emitting an electronic attack. The detecting unit reports its detection to all units using a given network when the data link is degraded or not operational.

D-3. Upon receipt of several messages, the source of enemy electronic attack can be determined by comparing lines of bearing from the different origins (triangulation).

D-4. See FM 6-99.2, page 83, for the format.

ELECTRONIC ATTACK REQUEST FORMAT

D-5. Electronic fires fall within three categories: preplanned, preplanned on-call, and immediate. Requesting airborne electronic attack support for ground operations is similar to requesting close air support. Requests for an electronic attack are sent via the normal joint air request process. Requesters use either a joint tactical air strike request or joint tactical air support request. (See FM 3-09.32 for a sample.) A theater-specific electronic attack request format may complement a joint tactical air strike request.

D-6. When submitting the request, the following information must be provided in the remarks section (section 8):

- Target location.
- Prioritized target description and jam frequencies.
- Time on target (window).
- Joint terminal attack controller.
- Jamming control authority call sign and frequency.
- Friendly force disposition (for example, troop movement route).
- Friendly frequency restrictions.
- Remarks.

ELECTRONIC WARFARE FREQUENCY DECONFLICTION MESSAGE

D-7. An EW frequency deconfliction message promulgates a list of protected, guarded, and taboo frequencies. This list allows friendly forces to use the frequency spectrum without adverse impact from friendly electronic attack. (See FM 6-99.2, page 86, for the format.)

Electronic Warfare Mission Summary

D-8. The EW mission summary summarizes significant EW missions and reports the status of offensive EW assets. EW and electronic-attack-capable surface and air units use it to provide information on EW operations. Service components use it to report significant events for subsequent analysis. (See FM 6-99.2, page 87, for the format.)

Electronic Warfare Requesting Tasking Message

D-9. Joint task force commanders use the electronic warfare requesting tasking message to task component commanders to perform EW operations in support of the joint EW plan and to support component EW operations. Component commanders use this message to request EW support from sources outside their command.

Joint Tactical Air Strike Request or Joint Tactical Air Support Request

D-10. Use a joint tactical air strike request or joint tactical air support request to request electronic attack. These requests require the information listed in paragraph D-6. Organizations without an automated capability submit these requests using DD Form 1972 (Joint Tactical Air Strike Request). See JP 3-09.3 and FM 3-09.32 for more information.

Joint Spectrum Interference Resolution

D-11. The joint spectrum interference resolution program replaced the DOD meaconing, intrusion, jamming, and interference program in June, 1992. Follow guidance in CJCSI 3320.02C to report incidents of spectrum interference.

Joint Restricted Frequency List

D-12. Operational, intelligence, and support elements use the joint restricted frequency list to identify the level of protection desired for various networks and frequencies. The list should be limited to the minimum number of frequencies necessary for friendly forces to accomplish objectives.

D-13. See Annex A to appendix B to JP 3-13.1 for the joint restricted frequency list format. The format is used by the joint automated communications-electronics operations instruction system. The format is unclassified but should show the proper classification of each paragraph when filled in. (See CJCSI 3320.01B and JP 3-13.1 for additional information.)

COUNTER-IMPROVISED-EXPLOSIVE-DEVICE ACTIVITIES

D-14. Certain reports and references are associated with counter-improvised-explosive-device activities. Most of these reports include information pertinent to counter-radio-controlled improvised-explosive-device EW activities. EW working groups have the responsibility to monitor these reports to assess planned counter-radio-controlled improvised-explosive-device EW operations and to support future operations. These reports typically use formats established in FM 6-99.2 modified to include improvised explosive device considerations and current operations. See GTA 90-10-046 for examples of reports and references applicable to counter-radio-controlled improvised-explosive-device EW operations.

Appendix E
Army and Joint Electronic Warfare Capabilities

This appendix provides information on Army and other Service electronic warfare capabilities. It is not an all-inclusive list. Due to the evolving nature of electronic warfare equipment and systems, this information is perishable and should be augmented, updated, and maintained by the unit electronic warfare officer.

ARMY

E-1. The Army is currently expanding its electronic warfare (EW) capability. It maintains several EW systems in its inventory. Currently, all units whose sole purpose is to conduct EW operations are assigned to 1st Information Operations Command. When requested, these capabilities are provided to combatant commands for employment at corps and lower echelons.

Counter-Radio-Controlled Improvised-Explosive-Device EW Systems

E-2. Counter-radio-controlled improvised-explosive-device EW systems form a family of electronic attack systems. Army forces use these systems to prevent improvised explosive device detonation by radio frequency energy. The Army maintains both a mounted and dismounted counter-radio-controlled improvised-explosive-device EW capability to protect personnel and equipment. For a detailed description of these systems, see appendix F.

Aircraft Survivability Equipment

E-3. Aircraft survivability equipment aims to reduce aircraft vulnerability, thus allowing aircrews to accomplish their immediate mission and survive. Army aviation maintains a suite of aircraft survivability equipment that provides protection against electronic attack. This protection can include radio frequency warning and countermeasures systems, a common missile warning system, information requirement countermeasures systems, and laser detection and countermeasure systems. For a detailed description of aircraft survivability equipment EW-related systems, see appendix F.

Intelligence Systems

E-4. The intelligence community maintains many systems that provide data for use in EW operations. Signals intelligence systems provide most of this required data. These assets are dual use. Usually the data collected is categorized as signals intelligence. It is maintained within sensitive compartmented information channels and governed by the National Security Agency/Central Security Service. The data sometimes support EW or, more specifically, electronic warfare support. Paragraphs E-5 through E-7 illustrate some intelligence systems that (when tasked) can provide electronic warfare support data to support electronic attack and electronic protection actions. For a detailed description of other intelligence and EW-support-related systems, see appendix F.

Guardrail Common Sensor

E-5. The Guardrail common sensor is a corps-level airborne signals intelligence collection and location system. (See figure E-1.) It provides tactical commanders with near real-time targeting information. Key features include the following: integrated communications intelligence and electronic intelligence reporting, enhanced signal classification and recognition, near real-time direction finding, precision emitter location, and an advanced integrated aircraft cockpit. Preplanned product improvements include frequency extension, computer-assisted online sensor management, upgraded data links, and the capability to exploit a wider range of signals. The Guardrail common sensor shares technology with the ground-based common sensor, airborne reconnaissance-low, and other joint systems.

Figure E-1. Guardrail common sensor

Aerial Common Sensor

E-6. The aerial common sensor is the Army's programmed airborne intelligence, surveillance, and reconnaissance system. (See figure E-2.) It will replace the current RC-7 airborne reconnaissance-low and Guardrail common sensor programs. The aerial common sensor uses the operational and technical legacies of the airborne reconnaissance-low and Guardrail common sensor systems as well as some technological improvements. This sensor will then provide a single, effective, and supportable multiple-intelligence system for the Army. The aerial common sensor will include a full multiple-intelligence capability, including carrying signals intelligence payloads, electro-optic and infrared sensors, radar payloads, and hyperspectral sensors.

Figure E-2. Aerial common sensor (concept)

Prophet

E-7. The Prophet system is the division, brigade combat team, and armored cavalry regiment principal ground tactical signals intelligence and EW system. (See figure E-3.) Prophet systems will also be assigned to the technical collection battalion of battlefield surveillance brigades. Prophet detects, identifies, and locates enemy electronic emitters. It provides enhanced situational awareness and actionable 24-hour information within the unit's area of operations. Prophet consists of a vehicular signals intelligence receiver mounted on a high mobility multipurpose wheeled vehicle, plus a dismounted-Soldier-portable version. The dismounted Soldier portable version is used for airborne insertion or early entry to support rapid reaction contingency and antiterrorist operations. Future Prophet systems are planned to include an electronic attack capability.

Figure E-3. Prophet (vehicle-mounted)

MARINE CORPS

E-8. The Marine Corps has two types of EW units: radio battalions (often called RADBNs), and Marine tactical EW squadrons (referred to as VMAQs). Paragraphs E-9 through E-24 discuss the units' missions, their primary tasks, and capabilities currently being employed. (For further information on the Marine Corps EW units and systems, see MCWP 2-22.)

RADIO BATTALION

E-9. Radio battalions are the Marine Corps' tactical level ground-based EW units. During operations, teams from radio battalions are most often attached to the command element (or senior headquarters) of Marine expeditionary units. Each radio battalion has the following mission, tasks, and equipment.

Mission and Tasks

E-10. The mission of the radio battalion is to provide communications security monitoring, tactical signals intelligence, EW, and special intelligence communication support to the Marine air-ground task force (MAGTF). The radio battalion's tasks include—

- Executing interception; radio direction finding; recording and analysis of communications and noncommunications signals; and signals intelligence processing, analysis, production, and reporting.
- Conducting EW against enemy or adversary communications.
- Helping protect MAGTF communications from enemy exploitation by conducting communications security monitoring, analysis, and reporting on friendly force communications.
- Providing special intelligence communications support and cryptographic guard (personnel and terminal equipment) in support of the MAGTF command element. Normally, the communications unit supporting the MAGTF command element provides communications connectivity for special intelligence communications.
- Providing task-organized detachments to MAGTFs with designated signals intelligence, EW, special intelligence communication, and other required capabilities.
- Exercising technical control and direction over MAGTF signals intelligence and EW operations.
- Providing radio reconnaissance teams with specialized insertion and extraction capabilities (such as combat rubber raiding craft, fast rope, rappel, helocast, and static-line parachute) for specified signals intelligence and limited electronic attack support during advance force, preassault, or deep postassault operations.
- Coordinating technical signals intelligence requirements and exchanging technical information and material with national, combatant command, joint, and other signals intelligence units.
- Providing intermediate, third, and fourth echelon maintenance of the radio battalion's signals intelligence and EW equipment.

Equipment

E-11. The following illustrate EW capabilities a radio battalion uses to accomplish the mission and perform the tasks in support of the MAGTF:

AN/ULQ-19(V)2 Electronic Attack Set

E-12. The AN/ULQ-19(V)2 electronic attack set allows operators to conduct spot or sweep jamming of single-channel voice or data signals. To provide the required jamming, the system must be employed and operated from a location with an unobstructed signal line of sight to the target enemy's communications transceiver.

AN/MLQ-36 Mobile Electronic Warfare Support System

E-13. The AN/MLQ-36 mobile electronic warfare support system provides a multifunctional capability that gives signals intelligence and EW operators limited armor protection. This equipment can provide signals intelligence and EW support to highly mobile mechanized and military operations in urban terrain where maneuver or armor protection is critical. This system is installed in a logistic variant of the Marine Corps's light armored vehicle. It consists of the following:

- Signals intercept system.
- Radio direction finding system.
- Electronic attack system.
- Secure communication system.
- Intercom system.

AN/MLQ-36A Mobile Electronic Warfare Support System (Product Improved)

E-14. The product-improved AN/MLQ-36A mobile electronic warfare support system (sometimes called the AN/MLQ-36A MEWSS PIP) is an advanced signals intelligence and EW system integrated into the Marine Corps's light armored vehicle. (See figure E-4.) This system replaces the equipment in the AN/MLQ-36.

E-15. The AN/AMLQ-36A has the following capabilities:

- Detect and evaluate enemy communications emissions.
- Detect and categorize enemy noncommunications emissions (such as battlefield radars).
- Determine lines of bearing.
- Degrade enemy tactical radio communications.

When mission-configured and working cooperatively with other AN/MLQ-36As, the system can provide precision location of battlefield emitters.

E-16. This system and its future enhancements will provide the capability to exploit new and sophisticated enemy electronic emissions and conduct electronic attack in support of existing and planned national, combatant command, fleet, and MAGTF signals intelligence and EW operations.

Figure E-4. AN/MLQ-36A mobile electronic warfare support system

Marine Tactical Electronic Warfare Squadron

E-17. Marine tactical electronic warfare squadrons are the Marine Corps's airborne tactical EW units. Each squadron has the following mission, tasks, and capabilities.

Mission and Tasks

E-18. The mission of the electronic warfare squadron is to provide EW support to the MAGTF and other designated forces. The squadron conducts tactical jamming to prevent, delay, or disrupt the enemy's ability to use the following kinds of radars: early warning, acquisition, fire or missile control, counterfire, and battlefield surveillance. Tactical jamming also denies and degrades enemy communication capabilities. The squadron conducts electronic surveillance operations to maintain electronic orders of battle. These include both selected emitter parameters and nonfriendly emitter locations. The squadron also provides threat warnings for friendly aircraft, ships, and ground units. Squadron tasks include—

- Providing airborne electronic attack and EW support to the aviation combat element and other designated operations by intercepting, recording, and jamming threat communications and noncommunications emitters.
- Processing, analyzing, and producing routine and time-sensitive electronic intelligence reports for updating and maintaining enemy electronic order of battle.

- Providing liaison personnel to higher staffs to assist in squadron employment planning.
- Providing an air EW liaison officer to the MAGTF EW coordination cell.
- Conducting electronic attack operations for electronic protection training of MAGTF units.

E-19. The squadron's EW division supports EA-6B Prowler tactical missions with intelligence, the tactical electronic reconnaissance processing and evaluation system (TERPES), and the joint mission planning system. All systems support premission planning and postmission processing of collected data, and production of pertinent intelligence reports. Working with squadron intelligence, these systems provide required electronic intelligence and electronic order of battle intelligence products to the aviation combat element, MAGTF, and other requesting agencies.

Equipment

E-20. Marine tactical electronic warfare squadrons maintain the following equipment:

- EA-6B Prowler.
- Joint mission planning system.
- Tactical electronic reconnaissance processing and evaluation system.

EA-6B Prowler

E-21. The EA-6B Prowler is a subsonic, all-weather, carrier-capable aircraft. (See figure E-5.) The crew consists of one pilot and three electronic countermeasure officers. The EA-6B has two primary missions. One is collecting and processing designated threat signals of interest for jamming and subsequent processing, analysis, and intelligence reporting. The other is employing the AGM-88 high-speed antiradiation missile against designated targets. The EA-6B's AN/ALQ-99 tactical jamming system incorporates receivers for the reception of emitted signals and external jamming pods for the transmission of energy to jam targeted radars (principally those associated with enemy air defense radars and associated command and control). In addition to the AN/ALQ-99, the EA-6B also employs the USQ-113 communications jammer to collect, record, and disrupt threat communications.

Figure E-5. EA-6B Prowler

Joint Mission Planning System

E-22. The joint mission planning system helps the EA-6B aircrew plan and optimize receivers, jammers, and high-speed antiradiation missiles. This system allows an operator to—

- Maintain area of operations emitter listings.
- Edit emitter parameters.
- Develop mission-specific geographic data and electronic order of battle to—
  - Tailor or create high-speed antiradiation missile direct attack libraries, or manually modify entries or new threat cards.
  - Plan target selection.
- Perform postflight mission analysis to—
  - Identify electronic emitters using various electronic parameter databases and electronic intelligence analytical techniques.
  - Localize emitters by coordinates with a certain circular error of probability for each site.
  - Correlate new information with existing data.
  - Gather postflight high-speed antiradiation missile information. This information includes aircraft launch parameters, predicted seeker footprint, and the onboard system detection of a targeted signal at impact.

AN/TSQ-90 Tactical Electronic Reconnaissance Processing and Evaluation System

E-23. The TERPES (AN/TSQ-90) is an air and land transportable, single-shelter electronic intelligence processing and correlation system. Each of the four Marine tactical electronic warfare squadrons includes a TERPES section.

E-24. A TERPES section consists of Marines, equipment, and software. The section identifies and locates enemy radar emitters from data collected by EA-6B aircraft and those received from other intelligence sources. It processes and disseminates EW data rapidly to MAGTF and other intelligence centers and provides mission planning and briefing support. Section support areas include operational support, intelligence analysis support, data fusion, fusion processing, and intelligence reporting. The section provides the following operational support:

- Translates machine-readable, airborne-collected, digital data into human- and machine-readable reports (such as paper, magnetic tape, secure voice, plots, and overlays).
- Receives and processes EA-6B mission tapes.
- Accepts, correlates, and identifies electronic emitter data from semiautomatic or automatic collection systems using various electronic parameter databases and various analysis techniques.
- Provides tactical jamming analysis.

AIR FORCE

E-25. The Air Force has two primary platforms that provide EW capability: the EC-130H Compass Call and RC-135V/W Rivet Joint. (For further information on Air Force EW equipment, see AFDD 2-5.1.)

EC-130H Compass Call

E-26. The EC-130H Compass Call is an airborne tactical weapon system. (See figure E-6.) Paragraphs E-27 through E-31 discuss the EC-130H missions, primary tasks, and capabilities.

Mission and Tasks

E-27. The EC-130H's mission is to disrupt enemy command and control information systems and limit the coordination essential for force management. The EC-130H's primary task is to employ offensive counterinformation and electronic attack capabilities in support of U.S. and multinational tactical air, surface, and special operations forces.

Figure E-6. EC-130H Compass Call

Capabilities

E-28. The EC-130H is designed to deny, degrade, and disrupt adversary command and control information systems. This includes denial and disruption of enemy surveillance radars; denial and disruption of hostile communications being used in support of enemy ground, air, or maritime operations; and denial and disruption of many modern commercial communication signals that an adversary might employ.

Compass Call During Operation Iraqi Freedom

During Operation Iraqi Freedom, much speculation appeared in the press about why Iraqi forces failed to ignite the oil facilities they had wired for destruction. During the coalition's seizure of Al Faw, Compass Call disrupted the Iraqi regime's control of its troops by jamming its communications. Instead of receiving orders to detonate the oil terminals, Iraqi troops heard only the ratcheting static of Compass Call jamming until coalition ground troops had secured the area. In addition to the conquest of the Al Faw Peninsula, successful military operations supported by Compass Call in Operation Iraqi Freedom included the seizure of four airfields; two successful prisoner of war rescues; and the ground offensive from Basrah to Nasariyah, Najaf, Baghdad, and Tikrit. In all these instances, Compass Call jamming prevented a trained, experienced enemy from coordinating actions against coalition forces.

"EC-130H Compass Call: A textbook example of Joint Force integration at its best", Electronic Warfare Working Group, U.S. House of Representatives, Issue Brief #17, 11 Mar 2004. (Available at http://www.house.gov/pitts/initiatives/ew/Library/Briefs/brief17.htm)

RC-135V/W Rivet Joint

E-29. Paragraphs E-30 through E-31 discuss the missions, primary tasks, and capabilities of the RC-135V platforms.

Mission and Tasks

E-30. The RC-135V/W Rivet Joint is a combatant-command-level surveillance asset that responds to national-level taskings. (See figure E-7.) Its mission is to support national consumers, combatant commanders, and combat forces with direct, near real-time reconnaissance information and electronic warfare support. It collects, analyzes, reports, and exploits information from enemy command and control information systems. During most contingencies, it deploys to the theater of operations with the airborne elements of the theater air control system.

RC-135V/W Rivet Joint, U.S. Air Force Fact Sheet, September, 2007.

Figure E-7. RC-135V/W Rivet Joint

Capabilities

E-31. The RC-135V/W is equipped with an extensive array of sophisticated intelligence gathering equipment that enables monitoring of enemy electronic activity. The aircraft is integrated into the theater air control system via data links and voice (as required). Refined intelligence data can be transferred from Rivet Joint to an Airborne Warning and Control System platform through the tactical digital information link. Alternatively, this data can be placed into intelligence channels via satellite and the tactical information broadcast service (a near real-time combatant command information broadcast). The aircraft has secure ultrahigh frequency, very high frequency, and high frequency (commonly known as UHF, VHF, and HF respectively) as well as satellite communications. It can be refueled in the air.

NAVY

E-32. The Navy's primary airborne EW platforms are the EA-6B Prowler and its planned replacement, the E/A-18G Growler. E/A-18G fielding is scheduled to begin in 2009 and is scheduled to replace the Navy's carrierborne EA-6B aircraft. The Navy also maintains both surface and subsurface EW shipboard systems for offensive and defensive missions in support of the fleet. (For further information on Navy missions and equipment, see NWP 3-13.)

EA-6B PROWLER

E-33. Paragraphs E-34 through E-39 discuss the missions, primary tasks, and capabilities of the Navy's EA-6B Prowler platforms. (See figure E-8.)

Figure E-8. Navy EA-6B Prowler

Mission and Tasks

E-34. The mission of the Navy's EA-6B Prowler is to ensure survivability of U.S. and multinational forces through suppression of enemy air defenses (using the radar-jamming AN/ALQ-99 tactical jamming system), lethal suppression (using the AGM-88 high-speed antiradiation missile), and communications jamming (using the USQ-113 radio countermeasures set). Prowlers have supported U.S. and multinational forces operating from various expeditionary sites throughout the world while maintaining full presence on all Navy aircraft carriers.

Capabilities

E-35. The Navy's EA-6B Prowlers are outfitted with either the improved capability II or improved capability III systems. The following lists the major capability upgrades these systems provide.

Improved Capability II

E-36. The improved capability II program was initiated in the 1980s. It was completed across the fleet of EA-6B aircraft (including U.S. Marine Corps aircraft) in the 1990s. The program incorporated incremental capability improvements that include communications, navigation, and computer interface upgrades; a high-speed antiradiation missile capability; and improved jamming pods. Several system interfaces were also upgraded in preparation for the improved capability III improvements.

Improved Capability III

E-37. The improved capability III program incorporates a highly evolved receiver system and provides upgraded EA-6B aircraft with increased signal detection, geolocation capability, a new selective reactive-jamming capability, and better reliability. High-speed antiradiation missile employment is also improved due to the speed of the receiver and its geolocation accuracy. Increased battlefield situational awareness of joint forces is also provided through Link-16. The improved capability III program provides a new ALQ-218 receiver system, integration of the USQ-113 and the multifunctional information distribution system (often called MIDS). This system incorporates Link-16 and various connectivity avionics into the Prowler. The major EW-related subsystems are the AN/ALQ-99 (V) tactical jamming countermeasures set and AN/USQ-113 (V) radio countermeasures set.

E-38. The AN/ALQ-99 (V) tactical jamming countermeasures set has upgraded receivers and processors to provide the following:

- Improved frequency coverage.
- Direction-of-arrival determination capability.
- Narrower frequency discrimination to support narrowband jamming.
- Enhanced interface with onboard systems.

E-39. The AN/USQ-113 (V) radio countermeasures set will enhance the aircraft's jamming capability through its integration with the tactical display system. This will enable the crew to display AN/USQ-113 communications jamming data as well as control AN/USQ-113 operations through the tactical display system.

E/A-18G GROWLER

E-40. The E/A-18G Growler is the Navy's replacement aircraft for the EA-6B Prowler. Paragraphs E-41 and E-42 discuss the missions, primary tasks, and capabilities of the Navy's E/A-18G Growler. (See figure E-9.) E/A-18G fielding began in 2008. The first operational E/A-18G deployment will occur in 2009, as the Navy begins to replace its carrierborne EA-6B aircraft.

Figure E-9. EA-18 Growler

Mission and Tasks

E-41. The EA-18G can detect, identify, locate, and suppress hostile emitters. It will provide enhanced connectivity to national, combatant command, and strike assets. Additionally, the EA-18G will provide organic accurate emitter targeting using on-board suppression weapons, such as the high-speed antiradiation missile.

Capabilities

E-42. The following is a list of the E/A-18G's general capabilities:

- Suppression of enemy air defenses. The EA-18G will counter enemy air defenses using both reactive and preemptive jamming techniques.
- Stand-off and escort jamming. The EA-18G will be highly effective in the traditional stand-off jamming mission, but with the speed and agility of a Super Hornet, it will also be effective in the escort role.
- Integrated air and ground airborne electronic attack. Enhanced situational awareness and uninterrupted communications will enable the EA-18G to achieve a higher degree of integration with ground operations than previously.
- Self-protect and time-critical strike support. With its active electronically scanned array radar, digital data links, and air-to-air missiles, the EA-18G will be able to protect itself and effectively identify and prosecute targets.
- Growth. High commonality with the F/A-18E and F/A-18F, nine available weapon stations, and modern avionics enable cost-effective synergistic growth, setting the stage for continuous capability enhancement.

E-43. The following is a list of the E/A-18G's airborne electronic attack capabilities:

- Entire spectrum. The EA-18G's ALQ-218 wideband receiver combined with the ALQ-99 tactical jamming system will be effective against any surface-to-air threat.
- Precision airborne electronic attack. Selective-reactive technology enables the EA-18G to rapidly sense and locate threats much more accurately than before. This improved accuracy enables greater concentration of energy against threats.
- Advanced communication countermeasures. Its modular communication countermeasure set enables the EA-18G to counter a wide range of communication systems and is readily adaptable to an ever changing threat spectrum.
- Interference cancellation system. This system dramatically enhances aircrew situational awareness by enabling uninterrupted communications during jamming operations.

CAPABILITIES SUMMARY

E-44. Table E-1 lists Army and joint EW capabilities. (Bold text indicates capabilities not described in the preceding paragraphs.) EW officers, noncommissioned officers, and supporting staff members should be familiar with these capabilities and how they can support Army operations. Additional information on the EW capabilities listed in table E-1 is found in the Web sites listed in table E-2, page E-12.

Table E-1. Army and joint electronic warfare capabilities

	Army	Air Force	Navy	Marine Corps
Airborne	RC-12 Guardrail	**EC-130J Commando Solo**	EA-6B Prowler	EA-6B Prowler
	airborne common sensor	EC-130H Compass Call	EA-18G Growler	
		RC-135V/W Rivet Joint	**EP-3E Aries II**	
		F-16CJ		
		E-8 JSTARS		
Unmanned aircraft system*	**RQ-5A/MQ-5B Hunter (Corps)**	**RQ-4A (Joint) Global Hawk**	**RQ-2 Pioneer**	
	RQ-7A/B Shadow (brigade)	**RQ-1L (Joint) Predator**	**MQ-8B Fire Scout Vertical Take-off**	
	MQ-1C/Sky Warrior (replacement for Hunter)	**RQ-11 Raven**	**Silver Fox**	
	MQ-8 Fire Scout			**RQ-11 Raven**
	RQ-11 Raven (battalion) Hand Launched			**Scan Eagle**
Ground	AN/MLQ-40 Prophet		CREW Systems (Joint)	AN/MLQ-36 MEWSS

Note: *Other Services may refer to unmanned aircraft systems as unmanned aerial systems or vehicles.
CREW counter radio-controlled improvised explosive device electronic warfare
MEWSS mobile electronic warfare support system
SOF special operations forces

Table E-2. Electronic warfare systems and platforms resources

Army platforms and systems http://www.sed.monmouth.army.mil/avionics/ http://www.sec.army.mil/secweb/fact_sheets/fact_sheets.php
Air Force platforms and systems http://www.af.mil/factsheets/factsheet.asp?fsID=182 http://www.airforce-technology.com/projects/#Unmanned_Aerial_Vehicles_(UAV_/_UCAV)
Navy systems platforms and systems http://acquisition.navy.mil/programs http://www.naval-technology.com/projects/ http://www.navy.mil/navydata/fact.asp
Marine Corps platforms and systems http://www.marcorsyscom.usmc.mil/sites/cins/INTEL/USMC%20CREW/index.html
Joint programs https://www.jieddo.dod.mil

Appendix F
Tools and Resources Related to Electronic Warfare

This appendix provides information on tools and reachback resources related to electronic warfare. Electronic warfare officers, noncommissioned officers, and supporting staff members should be familiar with these tools and resources and how to use them to support electronic warfare operations. Some tools and resources require an approved user account prior to being granted access.

ARMY REPROGRAMMING ANALYSIS TEAM

F-1. The Army Reprogramming Analysis Team (ARAT) supports tactical commanders. It provides timely reprogramming of any Army-supported software used for target acquisition, target engagement, measurement and signature intelligence, and vehicle and aircraft survivability (including that operated by other Services). The team provides software changes not readily possible by operator input to respond to rapid deployments or changes in the operational environment. See their Web site at https://ako.sec.army.mil/arat/index.html (Army Knowledge Online login required).

F-2. ARAT provides reprogramming support to counter-radio-controlled improvised-explosive-device (IED) electronic warfare (EW) (sometimes referred to as CREW), and other electronic systems.

F-3. The team is accessible via the Army Reprogramming Analysis Team's Warfighter Survivability Software Support Portal. A secure Internet protocol router network (SIPRNET) account is required to access the portal.

NATIONAL GROUND INTELLIGENCE CENTER

F-4. The National Ground Intelligence Center provides all-source analysis of the threat posed by IEDs produced and used by foreign terrorist and insurgent groups. The center supports U.S. forces during training, operational planning, deployment, and redeployment.

F-5. The center maintains a counter-IED targeting program (often called CITP) portal on its SIPRNET site. This portal provides information concerning IED activities and incidents as well as IED assessments.

ELECTRONIC ORDER OF BATTLE

F-6. An electronic order of battle details all known combinations of emitters and platforms in a particular area of responsibility. It consists of several reachback resources:

- National Security Agency-Electronic Intelligence Parameter Query.
- U.S. electromagnetic systems database.
- National Ground Intelligence System parametric information relational intelligence tool database.
- Military equipment parametrics and engineering database.

E-SPACE

F-7. E-Space is a Department of Defense (DOD) entity housed in the National Security Agency. It provides intelligence assistance (primarily signals intelligence) to deployed EW officers. E-Space is a reachback capability available to EW officers and spectrum managers that can be leveraged to provide all-source intelligence products and answers to requests for information and spectrum interference questions.

JOINT ELECTRONIC WARFARE CENTER

F-8. The Joint Electronic Warfare Center is DOD's only joint EW center of expertise. It provides EW subject matter expertise from a range of backgrounds, including people with current multi-Service operational experience. The center has a limited capability to perform modeling and simulation studies and EW red team support. It can deploy in a support role if approved by the U.S. Strategic Command.

JOINT IMPROVISED EXPLOSIVE DEVICE DEFEAT ORGANIZATION

F-9. The Joint Improvised Explosive Device Defeat Organization (known as JIEDDO) leads, advocates, and coordinates all DOD actions in support of efforts by combatant commanders and their joint task forces to defeat IEDs as weapon of strategic influence.

JOINT SPECTRUM CENTER

F-10. The Joint Spectrum Center ensures DOD effectively uses the electromagnetic spectrum in support of national security and military objectives. The center serves as DOD's center of excellence for electromagnetic spectrum management matters in support of the combatant commands, military departments, and DOD agencies in planning, acquisition, training, and operations.

F-11. The center maintains databases and provides data about friendly force command and control information system locational and technical characteristics. This information is used to plan electronic protection measures. These databases provide EW planners with information covering communication, radar, navigation, broadcast, identification, and EW systems operated by the DOD, other government agencies, and private businesses and organizations.

F-12. The center provides information on a quick-reaction basis in various formats and media to support EW planners and spectrum managers.

KNOWLEDGE AND INFORMATION FUSION EXCHANGE

F-13. The Knowledge and Information Fusion Exchange (sometimes called KnIFE) is a program sponsored by U.S. Joint Forces Command. It provides Soldiers with observations, insights, and lessons from operations around the world.

ADDITIONAL INFORMATION

F-14. Further information on the above tools and resources can be accessed through Army Knowledge Online. The links to these Web sites can be viewed by first accessing the "Army Operational Electronic Warfare Course" on Army Knowledge Online at http://www.us.army.mil/suite/page/400055 and then clicking on Folders >Links>EW links.

Glossary

SECTION I – ACRONYMS AND ABBREVIATIONS

ARAT	Army Reprogramming Analysis Team
C-3	operations directorate of a multinational (combined) staff
CJCSI	Chairman of the Joint Chiefs of Staff instruction
CJCSM	Chairman of the Joint Chiefs of Staff manual
COA	course of action
DD	Department of Defense (official forms only)
DOD	Department of Defense
DODI	Department of Defense Instruction
EW	electronic warfare
FM	field manual
FMI	field manual, interim
G-2	assistant chief of staff, intelligence
G-3	assistant chief of staff, operations
G-5	assistant chief of staff, plans
G-6	assistant chief of staff, signal
G-7	assistant chief of staff, information engagement
GTA	graphic training aid
HF	high frequency
Hz	hertz
IED	improvised explosive device
IO	information operations
IPB	intelligence preparation of the battlefield
ISR	intelligence, surveillance, and reconnaissance
J-2	intelligence directorate of a joint staff
J-3	operations directorate of a joint staff
J-5	plans directorate of a joint staff
J-6	communications system directorate of a joint staff
JFMO	Joint Frequency Management Office
JIOWC	Joint Information Operations Warfare Center
JP	joint publication
MAGTF	Marine air-ground task force
MC	Military Committee (NATO)
MCWP	Marine Corps warfighting publication
MDMP	military decisionmaking process

NATO	North Atlantic Treaty Organization
S-2	intelligence staff officer
S-3	operations staff officer
S-6	signal staff officer
S-7	information engagement staff officer
SIPRNET	SECRET Internet Protocol Router Network
STANAG	standardization agreement (NATO)
TERPES	tactical electronic reconnaissance processing and evaluation system
U.S.	United States

SECTION II – TERMS

communications security

(joint) The protection resulting from all measures designed to deny unauthorized persons information of value that might be derived from the possession and study of telecommunications, or to mislead unauthorized persons in their interpretation of the results of such possession and study. (JP 6-0)

computer network operations

(joint) Comprised of computer network attack, computer network defense, and related computer network exploitation enabling operations. (JP 3-13)

directed energy

(joint) An umbrella term covering technologies that relate to the production of a beam of concentrated electromagnetic energy or atomic or subatomic particles. (JP 3-13.1)

electromagnetic environment

(joint) The resulting product of the power and time distribution, in various frequency ranges, of the radiated or conducted electromagnetic emission levels that may be encountered by a military force, system, or platform when performing its assigned mission in its intended operational environment. It is the sum of the electromagnetic interference; electromagnetic pulse; hazards of electromagnetic radiation to personnel, ordnance, and volatile materials; and natural phenomena effects of lightning and precipitation static. (JP 3-13.1)

electromagnetic environmental effects

The impact of the electromagnetic environment upon the operational capability of military forces, equipment, systems, and platforms. It encompasses all electromagnetic disciplines, including electromagnetic compatibility and electromagnetic interference; electromagnetic vulnerability; electromagnetic pulse; electronic protection, hazards of electromagnetic radiation to personnel, ordnance, and volatile materials; and natural phenomena effects of lightning and precipitation static. (JP 3-13.1)

electromagnetic spectrum

(joint) The range of frequencies of electromagnetic radiation from zero to infinity. It is divided into 26 alphabetically designated bands. (JP 1-02)

electromagnetic vulnerability

(joint) The characteristics of a system that cause it to suffer a definite degradation (incapability to perform the designated mission) as a result of having been subjected to a certain level of electromagnetic environmental effects. (JP 1-02)

electronic attack

(joint) Division of electronic warfare involving the use of electromagnetic energy, directed energy, or antiradiation weapons to attack personnel, facilities, or equipment with the intent of degrading, neutralizing, or destroying enemy combat capability and is considered a form of fires. (JP 3-13.1)

electronic protection

(joint) Division of electronic warfare involving actions taken to protect personnel, facilities, and equipment from any effects of friendly or enemy use of the electromagnetic spectrum that degrade, neutralize or destroy friendly combat capability. (JP 3-13.1)

electronic warfare

(joint) Military action involving the use of electromagnetic and directed energy to control the electromagnetic spectrum or to attack the enemy. Electronic warfare consists of three divisions: electronic attack, electronic protection, and electronic warfare support. (JP 3-13.1)

electronic warfare support

(joint) Division of electronic warfare involving actions tasked by, or under direct control of, an operational commander to search for, intercept, identify, and locate or localize sources of intentional and unintentional radiated electromagnetic energy for the purpose of immediate threat recognition, targeting, planning, and conduct of future operations. (JP 3-13.1)

emission control

(joint) The selective and controlled use of electromagnetic, acoustic, or other emitters to optimize command and control capabilities while minimizing, for operations security: a. detection by enemy sensors; b. mutual interference among friendly systems; and/or c. enemy interference with the ability to execute a military deception plan. (JP 1-02)

joint restricted frequency list

(joint) A time a geographically-oriented listing of TABOO, PROTECTED, and GUARDED functions, nets, and frequencies. It should be limited to the minimum number of frequencies necessary for friendly forces to accomplish objectives. (JP 3-13.1)

working group

(Army) A temporary grouping of predetermined staff representatives who meet to coordinate and provide recommendations for a particular purpose or function. (FMI 5-0.1)

References

REQUIRED PUBLICATIONS

These documents must be available to intended users of this publication.

FM 1-02 (101-5-1). *Operational Terms and Graphics.* 21 September 2004.

JP 1-02. *Department of Defense Dictionary of Military and Associated Terms.* 12 April 2001. (As amended through 4 March 2008.)

JP 3-13.1. *Electronic Warfare.* 25 January 2007.

FM 3-0. *Operations.* 27 February 2008.

FM 5-0 (101-5). *Army Planning and Orders Production.* 20 January 2005.

FM 6-0. *Mission Command: Command and Control of Army Forces.* 11 August 2003.

FMI 5-0.1. *The Operations Process.* 31 March 2006.

RELATED PUBLICATIONS

These documents contain relevant supplemental information.

JOINT AND DEPARTMENT OF DEFENSE PUBLICATIONS

Most joint publications are available online: <http://www.dtic.mil/doctrine/jpcapstonepubs.htm.>

CJCSI 3320.01B *Electromagnetic Spectrum Use in Joint Military Operations.* 01 May 2005

CJCSI 3320.02C. Joint Spectrum Interference Resolution (JSIR). 27 January 2006 (with change 1 as of 25 February 2008).

CJCSI 3320.03A *Joint Communications Electronics Operation Instructions.* 11 June 2005.

CJCSM 3122.03C. *Joint Operation Planning and Execution System Volume II, Planning Formats and Guidance.* 17 August 2007.

CJCSM 3320.01B *Joint Operations in the Electromagnetic Battlespace.* 25 March 2006.

CJCSM 3320.02A *Joint Spectrum Interference Resolution (JSIR) Procedures.* 16 February 2006.

DODI 4650.01. Policy and Procedures for Management and Use of the Electromagnetic Spectrum. 09 January 2009.

JP 2-0. *Joint Intelligence.* 22 June 2007.

JP 2-01. *Joint and National Intelligence Support to Military Operations.* 07 October 2004.

JP 3-0. *Joint Operations.* 17 September 2006.

JP 3-09. *Joint Fire Support.* 13 November 2006.

JP 3-09.3. *Joint Tactics, Techniques, and Procedures for Close Air Support (CAS).* 03 September 2003.

JP 3-13. *Information Operations.* 13 February 2006.

JP 3-13.3. *Operations Security.* 29 June 2006.

JP 3-13.4 (JP 3-58). *Military Deception.* 13 July 2006.

JP 3-60. *Joint Targeting.* 13 April 2007.

JP 6-0. *Joint Communications System.* 20 March 2006.

ARMY PUBLICATIONS

Most Army doctrinal publications are available online: <http://www.army.mil/usapa/doctrine/Active_FM.html>.

FM 2-0 (34-1). *Intelligence.* 17 May 2004.

FM 3-09.32. JFIRE: *Multi-Service Tactics, Techniques, and Procedures for the Joint Application of Firepower.* 20 December 2007.

FM 3-13 (100-6). *Information Operations: Doctrine, Tactics, Techniques, and Procedures.* 28 November 2003.

FM 3-13.10 (3-51.1). *Multi-Service Tactics, Techniques, and Procedures for the Reprogramming of Electronic Warfare and Target Sensing Systems.* 22 January 2007.

FM 5-19 (100-14). *Composite Risk Management.* 21 August 2006.

FM 6-20-10. *Tactics, Techniques, and Procedures for the Targeting Process.* 8 May 1996.

FM 6-99.2 (101-5-2). *U.S. Army Report and Message Formats.* 30 April 2007.

FM 34-130. *Intelligence Preparation of the Battlefield.* 8 July 1994.

FMI 2-01. *Intelligence, Surveillance, and Reconnaissance (ISR) Synchronization.* 11 November 2008.

GTA 90-10-046. *MNC-I Counter IED Smart Book.* September 2008.

NATO PUBLICATIONS

Allied Joint Publication 3.6. *Allied Joint Electronic Warfare Doctrine.* December 2003.

MC 64. *NATO Electronic Warfare (EW) Policy.* 26 April 2004.

STANAG 5048 C3 (Edition 5). *The Minimum Scale of Convectivity for Communications and Information Systems for NATO Land Forces.* 16 February 2000.

OTHER PUBLICATIONS

AFDD 2-1.9. *Targeting.* 8 June 2006.

AFDD 2-5.1. *Electronic Warfare.* 5 November 2002.

Executive Order 12333. *United States Intelligence Activities.* 4 December 1981.

MCWP 2-22 (2-15.2). *Signals Intelligence.* 13 July 2004.

NWP 3-13. *Navy Information Operations.* June 2003.

SOURCES USED

Electronic Warfare Working Group, U.S. House of Representatives, Issue Brief #17. "Compass Call During Operation Iraqi Freedom." 11 March 2004. Available online at http://www.house.gov/pitts/initiatives/ew/Library/Briefs/brief17.htm.

PRESCRIBED FORMS

None

REFERENCED FORMS

DA Forms are available on the APD website (www.apd.army.mil). DD forms are available on the OSD website (www.dtic.mil/whs/directives/infomgt/forms/formsprogram.htm).

DA Form 2028. *Recommended Changes to Publications and Blank Forms.*

DD Form 1972. *Joint Tactical Air Strike Request.*

Index

Entries are by paragraph number unless specified otherwise.

Entries are by paragraph number unless specified otherwise.

Entries are by paragraph number unless specified otherwise.

Entries are by paragraph number unless specified otherwise.

Entries are by paragraph number unless specified otherwise.

Entries are by paragraph number unless specified otherwise.

Field Manual
No. 3-19.15

Headquarters
Department of the Army
Washington, DC, 18 April 2005

Contents

Distribution Restriction: Approved for public release; distribution is unlimited.

*This publication supersedes FM 19-15, 25 November 1985.

Page

Preface

Field Manual (FM) 3-19.15 addresses continental United States (CONUS) and outside continental United States (OCONUS) civil disturbance operations. Today, United States (US) forces are deployed on peacekeeping, peace enforcement, and humanitarian assistance operations worldwide. During these operations, US forces are often faced with unruly and violent crowds intent on disrupting peace and the ability of US forces to maintain peace. Worldwide instability coupled with increasing US military participation in peacekeeping and related operations requires that US forces have access to the most current doctrine and tactics, techniques, and procedures (TTP) necessary to quell riots and restore public order.

In addition to covering civil unrest doctrine for OCONUS operations, FM 3-19.15 addresses domestic unrest and the military role in providing assistance to civil authorities requesting it for civil disturbance operations. It provides the commander and his staff guidance for preparing and planning for such operations. The principles of civil disturbance operations, planning and training for such operations, and the TTP employed to control civil disturbances and neutralize special threats are discussed in this manual. It also addresses special planning and preparation that are needed to quell riots in confinement facilities are also discussed.

In the past, commanders were limited to the type of force they could apply to quell a riot. Riot batons, riot control agents, or lethal force were often used. Today, there is a wide array of nonlethal weapons (NLW) available to the commander that extends his use of force along the force continuum. This manual addresses the use of nonlethal (NL) and lethal forces when quelling a riot.

Appendix A complies with current Army directives, which state that the metric system will be incorporated into all new publications.

The proponent for this publication is HQ, TRADOC. Send comments and recommendations on Department of the Army (DA) Form 2028 (*Recommended Changes to Publications and Blank Forms*) directly to Commandant, US Army Military Police School (USAMPS), ATTN: ATSJ-DD, 401 MANSCEN Loop, Fort Leonard Wood, MO 65473-8929.

Unless this publication states otherwise, masculine nouns or pronouns do not refer exclusively to men.

Chapter 1

Operational Threats of the Civil Disturbance Environment

As the Cold War ended with the fall of the Berlin Wall and the collapse of Soviet communism in 1989, the Army was no longer concerned with containing the aggressive Soviet Union. The US Army began to transform because the battlefield on which it would fight had changed. Operations Just Cause, Joint Endeavor, Desert Shield, and Desert Storm have all been recognized as containing nontraditional battlefields. Following these operations, the Army was called upon to conduct peacekeeping, peace enforcement, and humanitarian assistance operations. These operations required US forces to protect refugees and humanitarian workers and keep hostile factions separate to enforce peace.

GENERAL CAUSES FOR CIVIL UNREST

1-1. In these modern times, demonstrations, civil unrest, public disorder, and riots happen for a number of reasons. Some of these reasons are economic hardships, social injustices, ethnic differences (leading to oppression), objections to world organizations or certain governments, political grievances, and terrorist acts. An event can be triggered by a single cause or a combination of causes. For example, operations in the Balkans involving civil unrest and riots were the result of ethnic hatred, a lack of civil authority, food shortages, a revolution, and religious-based fighting factions.

1-2. Demonstrations may range from simple, nonviolent protests that address specific issues, to events that turn into full-scale riots. Gatherings in protest are recognized rights of any person or group, regardless of where US forces may be operating. This fundamental right is protected under the *Constitution of the US*. During peacekeeping or peace enforcement operations, US forces should never violate basic civil or human rights. Most protesters are law-abiding citizens who intend that their protests be nonviolent, but some protest planners insist that the event involve some kind of violence. Often, in the media, protesters can gain sympathy for their cause by prompting authorities to take physical action against them. Violence is often the result of demonstrators beginning to conduct unlawful or criminal acts and authorities (who are responsible for the safety and welfare of all) enforcing the laws of the municipality, state, or nation. The depth of violence is determined by the willingness of demonstrators to display and voice their opinions in support of their cause.

1-3. Commanders must be aware of the possibility that some individuals or groups within an organized demonstration may have the intent to cause disruption, incite violence, destroy property, and provoke the authorities. The tactical situation and actions of the crowd should dictate control and enforcement options. Agitators and criminal infiltrators within the crowd can lead to the eruption of violence. Inciting a crowd to violence or a greater intensity of violence through the use of brutish enforcement tactics should be avoided. Publicity can be detrimental to authorities and beneficial to crowds because it can further their causes.

1-4. Community unrest results in urban conflicts that arise from highly emotional social and economic issues. Economically deprived inner city residents may feel that they are treated unjustly or ignored by people in power and authority. Tensions can build quickly in a community over a variety of issues, such as hunger, poor employment opportunities, inadequate community services, poor housing, and labor issues. Tension in these areas creates the potential for violence. When tensions are high, it takes a small (seemingly minor) incident, rumor, or act of injustice to ignite groups within a crowd to riot and act violently. This is particularly true if community relations with authorities are part of the problem.

1-5. Significant ethnic differences in a community can create an atmosphere of distrust, even hatred. Unrest among ethnic groups competing for jobs, living areas, and sparse essentials can cause an eruption of civil disorder and/or riots. As emotions run high, violence becomes likely.

1-6. Terrorist organizations may infiltrate groups within a demonstrating crowd. These terrorist groups may intend to embarrass their government or other governments. Terrorist infiltrators can be used to provoke crowds as a diversion, as part of a demonstration, or as cover for terrorist acts.

CROWD DEVELOPMENT

1-7. Crowds are a gathering of a multitude of individuals and small groups that have temporarily assembled in the same place. These small groups are usually comprised of friends, family members, or acquaintances that represent a group belief or cause. Individuals assume a sense of anonymity—they are viewed as just another face in the crowd. People in small groups are known only to companions in their group and to others in the gathering that have come from the same neighborhood or community. Commanders must consider how the individuals assembled and how they are interacting during the gathering process.

GATHERINGS

1-8. The assembly process of a gathering refers to the movement of people from different locations to a common location within a given period. This largely determines who participates. Creating a gathering from a crowd is a process with a beginning, middle, and end. As shown in *Figure 1-1* a crowd has an assembly process that leads to the gathering of the crowd, which is always followed by a dispersal process.

1-9. Gatherings are often assisted by the activities of individuals or groups with a specific agenda, such as yelling catchy slogans and cheers that everyone can easily pick up and join in on. Some groups are so well organized that they can prestage leaders to infiltrate a gathering. This creates unity, even inciting newcomers to join their cause. The first phase of the gathering is the assembly process. It can occur in one of two ways—impromptu or organized.

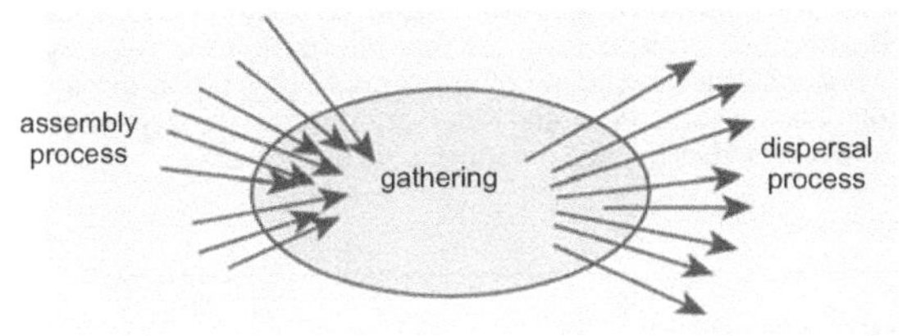

Figure 1-1. Crowd Building

IMPROMPTU GATHERINGS

1-10. Impromptu gatherings usually develop informally and are mostly done by word of mouth (one person to another or one group to another). Participants spread intelligence by telling one another when, where, and what is happening and inviting them to participate. An example of an impromptu gathering would be a gathering at a secured food distribution point after receiving information (by word of mouth) that a large truck carrying much-needed supplies and food is about to arrive. In this case, hunger would be the driving force causing the migration of people to the food distribution facility.

ORGANIZED GATHERINGS

1-11. Like an impromptu gathering, an organized gathering can also involve individuals and groups passing on information to one another. Passing intelligence of police activity and occurrences and on when and where events will take place helps organizers to prestage participants. Rarely is one represented group responsible for pulling together a gathering. Organized gatherings rely heavily on established groups that attract people to gather. Recent examples of these well-organized groups are anarchists, anti-globalization groups, and anti-free-enterprise groups. Groups representing extreme religious faiths and ethnic organizations have been common too. Some gatherings involve groups from rival, or even warring, factions.

1-12. Organized gathering processes rely more on centralized planning and organization. One or more of the groups offer organizers lists of individual names and groups, which they contact as potential participants. Modern technologies are available to these organizers through telephone banks, mailing lists, or e-mail addresses. In preparation for a long-term event, some group organizers have the means to provide logistical support of transportation, food, and water to participants.

> During the preelection period, Bosnian-Serbs (in a bid not to lose any of the territory they had won during the war) began to organize this fiercely contested city. Serbian mobs attacked US forces with bricks, rocks, metal bars, and firebombs. Organization for this altercation started days before, as many witnessed the prestaging of the items to be used in the attack. Wheelbarrows, 55-gallon metal drums, and other large containers were filled with the ammunition that would later be used to attack peacekeepers.
>
> Brcko, Bosnia
> September 1996

CROWD BUILDING

1-13. The second phase of the gathering process is the building of the crowd. Old concepts of crowds based on stereotypes are no longer true. Studies show that—

- Crowds are made of many entities, and all participants are not the same.
- Crowds are not made of isolated individuals but of a minority of individuals and a majority of small groups who may or may not be acquainted with each other.
- Groups and individuals in the crowd are not unanimous in their motivation.
- Groups and individuals in the crowd are usually not anonymous to each other, especially different groups gathered at the same place.
- Individuals within a crowd are not given to unique emotional displays; although known within a group, most individuals would not want to draw attention to themselves.
- Groups within a crowd will often act in unison in an attempt to achieve their agendas.
- Crowds are generally not unique or distinguished by violence or unlawful conduct.

DISPERSAL PROCESS OF A GATHERING

1-14. The last and final phase of any gathering of a crowd is the dispersal process. It involves the movement of people from the common location where they assembled to one or more alternate locations. The dispersal process ends the gathering of a crowd or at least begins its decline. The dispersal can occur on a routine, emergency, or coerced basis.

ROUTINE DISPERSAL

1-15. The routine dispersal may be specified in advance. It can also be included in the assembly instructions given by the organizers of an event.

EMERGENCY DISPERSAL

1-16. An emergency dispersal occurs when people evacuate an area in result of an unexpected crisis, such as a fire, explosion, bomb threat, or terrorist act. Individuals in such an emergency quickly recover from the initial shock. They keep their wits about them and improvise ways of extricating themselves and their companions from the dangerous situation. However, tragedies have occurred in some emergency dispersal situations. Therefore, it is important that forces on the scene are careful not to misinterpret what is occurring.

COERCION DISPERSAL

1-17. Coercion dispersal is caused by the use of force at some level. This is not necessarily the best way to force the dispersal of a crowd. The negotiated management of crowds is the preferred method and has proven to be highly successful in getting crowd organizers to police themselves, especially if the demonstration and/or protest leaders are available and willing to participate.

CROWD DYNAMICS

1-18. Understanding crowds and how individuals and groups form crowds through the gathering process is important because these issues apply to the dynamics of a crowd. Under most circumstances, gathered crowds are orderly and present little or no problems for authorities. Crowds consist of people who, although very motivated and passionate, are also subject to their own need for creature comforts. Rainy, cold, and nasty weather has a way of disheartening all but the few highly motivated and disciplined individuals. If problems exist, they usually fall into the following three categories:

- **Public disorder.** Public disorder is a basic breach of civic order. Individuals or small groups assembling have a tendency to disrupt the normal flow of things around them.
- **Public disturbance.** Public disturbance is designed to cause turmoil on top of the disruption. Individuals and groups assembling into a crowd begin chanting, yelling, singing, and voicing individual or collective opinions.
- **Riot.** A riot is a disturbance that turns violent. Assembled crowds become a mob that violently expresses itself by destroying property, assaulting others, and creating an extremely volatile environment.

1-19. Being part of a crowd of people has certain effects on different people. Each individual in a crowd is susceptible to behaving in a way that is contrary to their normal behavior. Some reasons for these behaviors are as follows:

- Crowds provide individuals with a sense of anonymity. With so many others, an individual realizes that he is just another face in the crowd, giving a sense of invulnerability.
- Crowd and individual behaviors are impersonal by nature. The "them-against-us" attitude affords those within the crowd the ability to freely (without hesitation or reservation) be verbally abusive, throw objects, or attack anyone who gets in the way.

- Crowds provide individuals with the idea that their moral responsibilities have shifted from themselves as a person to the crowd as a whole. Large numbers of people discourage individual behavior, and the urge to imitate others within the crowd is strong. Individuals look to others around them for cues of what to do next, disregarding their own background and training. Often, it is only the strong, well-disciplined person who can resist the prevailing behavior of a crowd.
- Crowd behavior influences the actions of both the disorderly individuals of the crowd and the authorities tasked to control them.

1-20. Individuals within a crowd are at times driven by deep-felt emotions. Emotional contagion is the most dramatic psychological factor of crowd dynamics. It provides the crowd with a temporary bond of psychological unity. Lasting long enough, this unity can push a simple organized crowd into a mob. Normal law and authority are rejected en masse under these conditions, increasing the potential for violence and panic to erupt.

1-21. Panic can erupt quickly, especially when crowds turn into mobs. Individuals within the mob can easily sense that their safety and well-being are at risk, putting them in a "fight-or-flight" mentality. Adding to the panic and confusion is the use of riot control agents (RCAs) by authorities in an attempt to gain control. Individuals in a mob, during the heat of confrontation, may attempt to leave and find that there are no escape routes and that roads are blocked. This can often lead to violent, physical attacks.

CROWD TYPES

1-22. Gaining an understanding of crowds and the dynamics that cause individuals or groups to join together in a crowd should be every leader's responsibility. Active, expressive, acquisitive, and hostile are not acceptable terms used to describe a crowd. They are the motives inferred from the actions the crowd takes. No gathering of a crowd is persistently or exclusively active, expressive, acquisitive, or hostile. For the sake of description, crowds can be identified as casual, sighting, agitated, or mob-like.

CASUAL CROWDS

1-23. Casual crowds are identified as individuals or small groups with nothing in common to bind them together. If they have an agenda, it is their own. They arrive separately and leave separately. Casual crowds are made up of individuals or small groups occupying the same common place, such as a shopping mall where these individuals or small groups meet.

SIGHTING CROWDS

1-24. Sighting crowds are similar to casual crowds with one additional element—an event. People migrate as a crowd to sporting events, are attracted to fires and accidents, and attend music concerts. Individuals or small groups gather at these events for the same purpose. It is the event and/or one's curiosity that compels a crowd to come together.

AGITATED CROWDS

1-25. Agitated crowds add responses that are based on the elements (people, space, and event). Individuals with strong emotional feelings within a crowd can quickly spread and infect the rest of the crowd. As more people within the crowd become emotionally involved, a sense of unity can develop, causing changes in the overall demeanor of the crowd. Yelling, screaming, crying, and profane name-calling are all associated with an agitated crowd.

MOB-LIKE CROWDS

1-26. Mobs have all the elements found in the first three types of crowds, with the addition of aggressive, physical, and sometimes violent actions. Under these conditions, individuals within a crowd will often say and do things they usually would not. Extreme acts of violence and property damage are often a part of mob activities. Mobs consist of (or involve) the elements of people and groups being mixed together and becoming fluid.

CROWD TACTICS

1-27. During a public disorder or disturbance, individuals and small groups within a crowd use any number of tactics to resist authority and disrupt and add turmoil in order to achieve their goals. These tactics can be unplanned or planned and violent or nonviolent. The more organized and purposeful a crowd becomes, the more likely a tactic will be used.

NONVIOLENT BEHAVIOR

1-28. As explained earlier in this chapter, most gatherings of individuals and small groups into a crowd do not involves violent behavior. A public disorder or disturbance usually involves some harmless name-calling, demonstrations to express views, corporate yelling and chanting, and even singing and dancing. In some instances, there may be property damage and the erection of makeshift barricades. All four types of crowds can be nonviolent, but mobs and agitated crowds have the greatest tendency to turn violent. Nonviolent actions of a crowd are disruptive because they are in direct conflict with what authorities want them to do, such as refusing to leave when directed, locking arms, and sitting in front of or around areas and buildings that the authorities are attempting to clear.

VIOLENT BEHAVIOR

1-29. A crowd that becomes a mob can be very violent and destructive. Although some nonviolent activities occur, violent crowds strike out physically at bystanders or others in the crowd, destroying both private and government property, setting fires, and employing bombs (in extreme cases). The only limitations for violent crowds are their own imaginations, the training of their leaders, and the materials readily available.

1-30. Violent crowds may erect barricades and physical barriers to impede the movement of authorities and prevent them from entering certain areas and/or buildings and to hide their activities. Common materials used for barricades include vehicles, felled trees, furniture, felled light poles, fencing materials, or anything else on hand and readily available.

1-31. Setting fires and preventing firefighters access to the structure by blocking it is another violent crowd tactic. Often, fires are set to create confusion in authorities or as a diversion to another activity.

RIOTS

1-32. A riot is one or more groups or individuals who are part of a larger crowd that involves threats of violence against persons or property. In some cases, a crowd will continue to gather until it evolves into a riot.

1-33. Riots vary considerably in both targets and players. A communal riot, for instance, deals with deep-seated ethnic, religious, and language differences. Commodity riots involve an attack on property by acts of vandalism, looting, and arson. Protest riots, such as the riot around the World Trade Organization (WTO) Assembly in Seattle, Washington (30 November 1999), illustrated individuals and groups aggressively and sometimes violently acting out or voicing their opposition to the assembly. The Democratic Convention in Chicago, Illinois, in 1968 reflects a riot that directly targeted police and authority in general. Celebration riots occur across the US as a result of home team victories in sporting events, among other reasons. Celebrating crowds look to make the moment more memorable through raucous acts that demonstrate their joy or happiness, for example, the riots that took place in Chicago in 1992 as a result of the Chicago Bulls winning the National Basketball Association (NBA) Championship.

1-34. Commanders and leaders must be aware that highly organized groups of protesters and/or demonstrators have developed tactics to disrupt the control force. These crowd tactics were published in handbooks for communist organizers during the Cold War, and today the Internet contains sites that are devoted to sharing these tactics with the general public and various protest groups. Although these tactics are somewhat outdated, commanders and leaders must be aware of them and their potential consequences. Many of these groups, such as anarchists, often mock the tactics employed against large crowds; for example, they may identify the police as "Darth Vader Cops" when the control force response is in full riot control gear. Commanders must be aware that well-organized crowds may attempt to engage the control force, surround it, and overpower it by sheer numbers. To avoid this, commanders must do a detailed terrain analysis (intelligence preparation of the battlefield [IPB]) of the area to include all approach and exit routes in the area. The control force flanks have to be protected to avoid being enveloped by the mob. To avoid being enveloped, the on-site commander should keep the crowd at a comfortable distance from the control force formation, which is usually the lethal zone of his NL munitions. Establishing predetermined rally points for the control force is critical in the event that this type of tactic is used by the crowd.

TACTICS USED TO DEFEAT AUTHORITIES

1-35. Organized mobs will try to defeat the control force by employing several different types of tactics. These tactics include the following:

- Constructing barricades.
- Using Molotov cocktails, smoke grenades, rocks, and slingshots.
- Feinting and flanking actions.

BARRICADES

1-36. By constructing barricades, the rioters are trying to protect themselves from assault by the control force. If the control force does advance toward the barricade, rioters hope that the barricade will split the control force and leave it vulnerable to actions by the mob.

MOLOTOV COCKTAILS, SMOKE GRENADES, ROCKS, AND SLINGSHOTS

1-37. Angry mobs will often attempt to disrupt the control force by throwing rocks and other projectiles. As the situation deteriorates, the mob may escalate the violence by using a battery of slingshots that will pellet the control force with a barrage of projectiles ranging from small rocks to marbles. They may also use smoke grenades (homemade or store-bought) to mask their movements. Finally, Molotov cocktails are used against personnel, employed vehicles, and portions of the control force. The mob will attempt to gain a position above an armored vehicle to enable them to drop a Molotov cocktail into an open hatch.

FEINTING AND FLANKING ACTIONS

1-38. Organized mobs may attempt to disrupt the movement of the control force by feinting an assault. When the control force moves to blunt the assault, the mob will assault the exposed flanks in an attempt to split the control force and envelop a portion of it.

WEAPONS

1-39. Mobs will often use various types of weapons against authorities to achieve their agendas and goals. It is extremely important that leaders train soldiers to recognize possible threats of the various weapons used by crowds or mobs, enabling the soldiers to react with the appropriate minimum force.

VERBAL ABUSE

1-40. In almost every instance of a civil disturbance or riot, verbal abuse will be an aggressive tool. Obscene language, racial remarks, taunts, ridicules, and jeers should always be expected. It is apparent that the purpose for using verbal abuse is to anger, demoralize, and provoke a physical response. Undisciplined, untrained soldiers who face such an attack could cause the situation to escalate. Just one provoked action of a soldier could be interpreted as an act of brutality by the media.

PERCEIVED INNOCENT PEOPLE OR THE WEAK

1-41. Women, children, and the elderly have often been used as pawns in civil disturbance operations. On many occasions, the innocent and the weak are moved to the front of the crowd and used as a barrier. With the innocent and the weak directly facing authorities, aggressive and violence-prone individuals or groups behind them attempt to provoke authorities to react. Often, this tactic is to exploit the situation in the eyes of the media for propaganda value. Despite the perception of sometimes being weaker, women, children, and the elderly can be just as committed to the insurgency as those behind them.

BARRICADES

1-42. Barricades are made of anything large or heavy enough to impede or prevent the movement of authorities. Trees, vehicles, fires, and furniture are all examples of items used to erect a barricade.

THROWN OBJECTS

1-43. In many cases, a crowd will throw anything in their reach that can be picked or picked up. Rotten fruit and/or vegetables, rocks, bricks, bottles, sticks, and pieces of lumber are most often used because they are readily available. More sophisticated types of thrown objects are improvised bombs, such as petrol bombs or tennis balls with nails sticking out of them. Never discount that these individuals and groups could acquire hand-thrown explosives, such as sticks of dynamite or grenades.

VEHICLES AND OTHER OBJECTS

1-44. Vehicles and other objects can be placed on higher ground (atop a hill or in a building) to disrupt civil disturbance operations. For example, a vehicle or tire can be set on fire and rolled down a hill, carts or barrels filled with hazardous or flammable materials can be pushed down a hill, or burning furniture and petrol bombs can be thrown from two-story or higher buildings or toward a formation of soldiers.

FIRE

1-45. Rioters have set buildings, vehicles, and other structures on fire to block the advances of authorities and to create confusion and diversions. If property damage is the goal of a crowd, fire is an effective tool. A sniper, for example, may set a building on fire to divert attention or provide maximum target possibilities. Other tactics used, particularly in the Balkans, were to flood the lower floors or basements of structures with flammable liquids or gas and then ignite it in an opportune moment.

FIREARMS

1-46. Sniping or massing fire from within the crowd, in buildings, or other covered positions can (at times) be effective tools for terrorists using a crowd to cover their acts. Firearms are anything that fires a projectile toward a target of opportunity.

EXPLOSIVES

1-47. A bomber's imagination is the only limiting factor in the use of explosives. Explosives can be used as simple diversions to block the advances of formations or provide an escape for rioters. The worst use of explosives is to purposely kill, injure, or demoralize authorities that are protecting and assuring the safety of everyone.

Chapter 2

Control Force Operations

Information is the key to developing civil disturbance plans. One must know who the demonstrators are; when, where, and why they are demonstrating; what their capabilities are; and what their possible course of action (COA) is. A commander's need for current, valid information cannot be overemphasized. The commander must learn as much as he can about the participants (their motivations, strategies, tactics, targets, and dedication). The more knowledge a commander has about the participants, the better equipped he is to counter their actions. He needs sound information to decide how to best use his available resources.

CIVIL DISTURBANCE PLANNING CONSIDERATIONS

2-1. The side that possesses better information and uses that information to gain understanding has a major advantage over its opponent. Forces that have this advantage can use it to bring about changes in attitudes, decisions, and actions in assembled demonstrators. In planning for crowd control or civil disturbance operations, planners must decide what data is needed to develop threat assessments. This decision is based on a multitude of information sources.

STANDARDIZED POSTURES IN COMPANY TACTICAL STANDING OPERATING PROCEDURES

2-2. Within squads, platoons, and companies, equipment may be increased or decreased, as needed. Some examples are as follows:

- Choose the M9 pistol for extraction and apprehension teams. The use of a long weapon (for example, an M16 with an M203, an M4 with an M203, or a 12-gauge shotgun) with NL munitions capability is also recommended, especially for overwatch personnel.
- Add nonstandard weapons such as shotguns for greater NL capabilities.

NOTE: The shotgun is used to protect the M203 gunner as he reloads.

- Add NL munitions to existing organic weapons systems, such as the M203.
- Arm soldiers in the front line of the formation with their standard weapon. If the weapon is a long weapon, it should be carried across the back from left to right or vice versa, with the butt up and muzzle down. Ensure that the weapon is cleared and that the magazine is in the appropriate ammunition pouch.

- Balance the mix of weapons and munitions according to the mission, enemy, terrain, troops, time available, and civilian considerations (METT-TC).
- Establish command relations and determine who has or gives the authority to fire NL munitions.
- Maximize the distance and barriers between the crowd and control formations. Use NL munitions to create a standoff distance.
- Maintain a lethal overwatch of the control force at all times.
- Maintain a reserve force at all times to reinforce the control force, as needed.
- Create NLW range cards for static positions.
- Consider environmental conditions and their effect on the performance of NL munitions. Proper storage, periodic inspections, and the rotation of NL munitions are critical to maintaining the effectiveness and viability of NL munitions.

WEAPON AND AMMUNITION CONFIGURATIONS

2-3. Individuals designated as NL shooters must have the means to transition to lethal rounds, if required. Lethal rounds are carried separately from NL rounds so that the shooter will not confuse them in the heat of confrontation. However, in the heat of confrontation where a lethal option is necessary, NL rounds can be used lethally by adjusting the point of aim to a vulnerable part of the body and closing in on the distance to the target.

2-4. Squad leaders should designate NL shooters within their squad. Ideally, the squad should not change its organization to accommodate NLW.

2-5. Commanders should not dispatch NL patrols. They should plan a combat or security patrol with NL capabilities. Soldiers are never sent in harm's way without lethal protection. NLW are only considered additional tools for the mission and not a mission in itself. There is no such thing as a NL mission.

2-6. Soldiers manning a static position should have NL capabilities depending on METT-TC. Shotguns and the M203 work well at static positions.

2-7. Recovery, apprehension, and/or extraction teams should be established before deployment. Team members should be equipped with personal protective equipment (PPE), an M9, NL munitions and weapons, and Flex-Cufs®. They should have some training in open-hand control, pain compliance, and handcuffs and/or Flex-Cufs.

2-8. Crowd control formations (see *Chapter 6*) should be well trained and well rehearsed. Rapid, coordinated movements of a well-trained and well-rehearsed control force can often be a strong enough deterrent. A lethal overwatch marksman always covers the control force.

2-9. During a NL engagement, the use of designated marksmen (DM) provides confidence and safety to those facing a riot. The DM in an overwatch position scans the crowd to identify threats and designate personnel for recovery and lethal rounds firing (if a lethal threat is presented). They are ideally suited for flank security and countersniper operations.

NONLEHTAL PLANNING

2-10. The fundamental principles of mission planning are not changed by the incorporation of NLW into a plan. NLW are intended to be operational-enhancement tools. The first concern of a commander is the successful completion of the mission. NL munitions provide a commander with a wider range of response options, but these munitions are not a replacement for lethal capabilities. Commanders must always be capable of answering a lethal attack with a lethal response.

2-11. When a commander commits his soldiers and equipment to a crowd control situation, he commits his forces with the additional tools of NL capabilities. This is an addition to the force continuum that the force commander now has available to him. Commanders that properly employ NL munitions and weapons have a tactical advantage over those who rely completely on lethal means.

2-12. When developing a tactical standing operating procedure (TSOP) (see *Chapter 6*), units should consider the following:

- Crowd control formations.
- Extraction teams.
- Apprehension teams.
- Lethal overwatch marksmen and/or observer teams.
- Reserve and/or security forces.

MILITARY WORKING DOG TEAMS

2-13. Military working dog (MWD) teams may be employed with a control force formation as a method of increasing crowd apprehension about approaching or engaging the formation. The teams should be in the rear of the formation in plain sight of the crowd, but in front of the command element and the M33A1 squad RCA disperser. The MWD teams work back and forth behind the formation as an intimidation measure. The presence of the MWD may produce a profound psychological effect on the crowd. These teams may also be used to help control individuals who have been captured by the recovery and apprehension teams.

CAUTION

Do not unleash an MWD on a crowd.

COMBAT CAMERAMEN

2-14. Video and still cameramen should make a photographic record of the individuals in the crowd who are leaders and instigators. Events must be documented to hold personnel, factions, gangs, or groups accountable for acts that violate law, destroy property, or cause physical harm. Electronically recording events aids in the prosecution of such cases and eliminates the sense of anonymity that people in large crowds often feel.

INFORMATION OPERATIONS

2-15. Information superiority helps forces to anticipate problems and requirements. It allows commanders to control situations earlier and with less force, creating the conditions necessary to achieve the optimal end state. Public affairs, psychological operations (PSYOP), and civil military operations are activities that will allow the commander to control situations earlier and with less force.

2-16. Information is available from a multitude of sources. A diversity of sources is the best approach because it prevents biased perspectives. Primary sources are as follows:

- Open sources.
  - Libraries.
  - Newspapers and news periodicals.
  - Radio and television.
  - Internet.
- Law enforcement sources.
  - Local law enforcement agencies.
  - National law enforcement agencies.
- Military sources.
  - Department of Defense (DOD) intelligence community (most restrictive source).
  - Local military intelligence (MI) field offices.

2-17. Police intelligence operations (PIO) is one of the five military police functions. This function is a process of actively and passively collecting information that is of a police, criminal, or combat nature. As military police perform the other four functions of maneuver and mobility support (MS), area security (AS), law and order (L&O), and internment and resettlement (I/R), they are gathering information that supports, enhances, and contributes to the commander's protection program, situational awareness, and battlefield visualization by portraying relevant threat information that may affect operational and tactical environments.

2-18. IPB is a continuous process for analyzing the threat and the environment of a specific geographic area. During the IPB process, the Intelligence Officer (US Army) (S2) or Assistant Chief of Staff, G2 (Intelligence) (G2) uses all available databases, intelligence sources and products, and related MI discipline to analyze the threat and the environment. The PIO function supports this process by providing the S2 with collected police, criminal, and combat information that can directly and significantly contribute to the success of the MI effort. In addition to combat information, the PIO function provides additional information on possible criminal threats and COAs. This is intended to support the S2 IPB process and can be used by the commander to upgrade force protection. (See *FM 19-10.*)

THREAT ANALYSIS

2-19. Crowd control requires its own thought process. Emphasis should be on prevention rather than confrontation. In combat, military forces are taught to fight and eliminate threats. In crowd control, military forces must deal with noncombatants that have internationally recognized rights. These rights must be respected while maintaining public order. This is an issue that law enforcement agencies (LEAs) have been struggling with for years.

2-20. Dealing with crowd control incidents is a matter of using some basic guidelines. It also includes asking the right questions in a logical manner so that key issues are not omitted. The commander must use these guidelines in context with METT-TC and the location of the incident (CONUS or OCONUS).

PREINCIDENT PLANNING

2-21. This phase of planning begins before the incident and is initiated at the operational level with guidance from the strategic level. The planning includes guidance on crowd control and addresses responsibilities, training, organization, operating procedures, use of force (CONUS), and/or rules of engagement (OCONUS). The most difficult and productive decisions are those made in the preincident planning process.

Avoid Confrontation

2-22. Crowd situations are highly unpredictable, but one thing seems certain—confrontation will likely cause crowd resistance. When pushed, people tend to resist opposition to the realization of their purposes.

Focus on Prevention

2-23. Planning should key in on the prevention of unfavorable outcomes. Experience has shown some LEAs attempting to help crowds accomplish their goals within the law that have been beneficial and even led to conceding some violations for the purpose of avoiding confrontation. However, LEAs maintain a law enforcement presence, which signifies social restraint. LEAs also decide when and where they will not compromise and the amount of force to use.

Define Goals

2-24. When defining a goal, deciding what must be accomplished is the first step. Defining the goal is fundamental. However, actually working toward and accomplishing the goal is easier said than done, as the process can drive the situation. Commanders and leaders must be aware of this. The military force must focus on what they are trying to accomplish.

BACKGROUND INFORMATION AND INTELLIGENCE

2-25. Seek to know as much as possible about social protest groups (within the limits of the law, see *Appendix B*) before an incident. This will provide insight into the organization and its functions and provide a warning as to what to expect. When you know as much about these organizations as possible, then it is possible to anticipate their next move. See *Appendix C*.

2-26. Continually assess what is to be accomplished against what others are trying to accomplish. Some groups may not have goals that conflict with those of US forces, but their actions may. Other groups may have goals and activities that do conflict. In either case, the assessment of group goals compared to the goals of US forces helps to understand and avoid potential conflicts.

CIVIL DISTURBANCE OPERATIONS PREPARATION

2-27. Counterdemonstration workgroups are a mixture of traditional and nontraditional staff proponents brought together to coordinate the resources to be employed by task force (TF) units before a potential civil disturbance. Meetings of the counterdemonstration workgroup are held on a regular basis (generally weekly) and are chaired by the Assistant Chief of Staff G3 (Operations and Plans) (G3). The following staff proponents may be involved, and others may be involved as needed:

- Public affairs office (PAO).
- Joint military commission (JMC), if organized.
- Provost marshal office (PMO).
- PSYOP.
- G2, Assistant Chief of Staff G5 (Civil Affairs) (G5), Assistant Chief of Staff G6 (Signal) (G6).
- Civil affairs (CA).
- Army airspace command and control (A2C2).
- Chaplain.
- Engineer.
- Surgeon.
- Fire support element (FSE).

RELATIONSHIPS DEVELOPMENT

2-28. Working relationships between commanders and protest group leaders are increasingly seen as the best means for preventing bad outcomes in crowd situations. This is called the negotiated management model of crowd control. LEAs in large metropolitan cities in the US and Europe practice it.

2-29. Open dialog helps develop working relationships between commanders and protest group leaders, providing an opportunity to communicate clearly. Talking allows group leaders to tell authorities and tactical commanders what they want to accomplish. It also allows authorities and commanders to tell group leaders what they are prepared to do and how they might respond to certain crowd behaviors. Such communications can do much to resolve issues and prevent violence.

2-30. Commanders may not be able to talk to all leaders before a crowd assembles. Some groups do not have recognized leaders and are ad hoc organizations. Other groups may have several leaders, but only some of those leaders will negotiate. In such cases, meetings with some leaders may tell you who the other leaders are and provide critical information. Negotiations may also encourage more moderate leaders to do things that will support the commander.

2-31. Initiate communication with group leaders to work out issues before an event. Commanders should make a concentrated effort to maintain a working relationship with the leaders of protest groups.

2-32. Communication also means persuasion. Commanders should make a concerted effort to win over demonstration leaders. Commanders may tell group leaders that they want to help them complete their mission in a safe manner. They should encourage demonstrators to protest in an acceptable way and, if possible, offer favors to get them to do just that.

NOTE: Commanders must only offer what they are willing and able to deliver.

2-33. Several rules apply in negotiations, and credibility is key. Only communicate necessary information and those actions that authorities and commanders intend to do. Negotiations are made from a position of strength, and by negotiating an agreement may be reached. However, contingencies should be developed in the event that the agreement is violated.

2-34. Working relationships between commanders and protest group leaders often result in protest groups policing themselves. This is one of the basic premises of negotiated management. Protesters are more likely to listen to their own leaders, as opposed to listening to US forces. Protest organizers are likely to buy into a cooperative effort and agree to ground rules set during preprotest negotiations. Many protest leaders seek to stay on the right side of the law.

ACCEPTABILITY

2-35. Deployed US forces will find themselves engaged in crowd control operations under difficult circumstances. The host nation (HN) may have groups that do not accept the presence of US forces, which will make imposing order and the protection of citizens difficult at best. US forces will be under intense media and political scrutiny (an environment much like what LEAs operate in on a day-to-day basis).

2-36. Winning in this environment is not like winning in combat. US forces may appear to be invincible and formidable, but they risk being portrayed as oppressors. Thus, US forces can lose by appearing to win. Groups that perceive themselves as oppressed will readily seek victimhood in an effort to gain the support of public opinion. Winning in this environment is about seizing and holding the moral high ground. US forces must maintain the authority and legitimacy of what they are doing.

2-37. Projecting a favorable image will require outreach to local leaders and citizens. It will also require developing a relationship with the media. For example, the Los Angeles Sheriffs Department often invites the media to accompany their tactical commander during crowd control situations. This shows that they have nothing to hide. It also provides an opportunity for individuals to see the commander's side of an event.

2-38. In this environment, commanders must consider how actions will play among several audiences—local, allied, US, and international. The media effect is inescapable. In some cases, it will be difficult to accommodate all of these audiences. However, the most important audience will be the US public, which is key to the continued support of US forces.

CROWD ASSESSMENT QUESTIONS

2-39. Assessing crowds requires its own set of questions. These questions should be answered before a crowd assembles or as quickly as possible if a gathering occurs without notification. In some cases, all the questions may not be answered. These questions represent a logical way to think about crowds. The questions are as follows:

- **Who are they? What is the overarching identity of the crowd? Are they strikers, ethnic factions, or social protesters? Do they identify themselves as strikers, ethnic groups, religious factions, or protesters against some perceived social injustice?** Understanding who they are will indicate what they may do. It may be possible to determine identities and goals from advance assembling instructions, leaflets distributed to bystanders, placards and banners, and chants and songs.
- **What are their goals?** What the group wants to accomplish by assembling could determine the extent to which they can be accommodated once they have assembled. They may only seek recognition for their cause (being seen and heard). If so, this goal is usually easy to accommodate. However, some groups may have more demanding goals, for example the demonstrators in Seattle who sought to stop the WTO. Goals that cannot be accommodated make confrontation very likely.
- **What is the composition of the crowd and are there any known factions?** Seattle demonstrations against the WTO were comprised of groups that were protesting environmental issues, wages, and child labor laws. Differing goals and the resulting friction were evident between such organizations as the Ruckus Society and the American Federation of Labor-Congress of Industrial Organizations (AFL-CIO). Factions within a crowd represent threats and opportunities.
- **What are they capable of doing?** Protest groups often claim that they will assemble large numbers of people to produce some disruptive action. However, there is often a difference between the claim and reality. An organization may claim that it can mass a 100,000 people, but in reality can only get 40,000. Organizers exaggerate for the following reasons: they want to boost the morale of their own people and they want the media to report that they have strength in numbers. Studying the past activities of a group may provide indications of what they are capable of doing in the future.
- **What are their traditional behaviors or cultural repertoires?** What people do during protests is not universal. It varies with the group and the culture. Social protest organizations and striking unions will carry placards and banners. Other groups will protest in a more quiet way, like the 5,000 women in Sarajevo whose standard Sunday behavior was to sit and block traffic. They were protesting the loss of male relatives in the Bosnian war. Understanding the goals of

the protesters can be helpful in deciding how to respond to their behaviors.

- **When and where will they assemble?** Every protest organizer has a time and place for assembling and perhaps a destination for the crowd to move toward. If the organizer attempts to mobilize large numbers of participants, the time and place for assembling and dispersing must be made known in the instructions. This information may be stated in the mobilization instructions or disseminated by an informal network, such as word of mouth. Such impromptu networks in densely populated areas can enable rapid assembling.
- **Where will they go?** Many crowds have destinations. Organizations may march a specified distance to ensure that their cause gets sufficient attention. Commanders need to know the route to minimize disruptions to the rest of the community. In Los Angeles, LEAs attempt to reroute traffic and prevent congestion caused by protest marches. It is also necessary to provide security along the route to prevent counterdemonstrators from confronting the marchers. This will prevent an even greater problem for the community.
- **What are the possible targets of violence?** Riots, in particular, may focus on target facilities. In the 1992 Los Angeles riots, gun stores were major targets. In developing countries, targets may be more basic. In Somalia, throngs of people stormed food supply facilities. In Haiti, riots broke out over trash dumps where US forces had discarded the remnants of meals, ready to eat (MREs).
- **What is the worst-case scenario?** The worst-case scenario must be recognized for the sake of avoidance. This may be when a peaceful crowd degenerates into a violent riot.
- **When and where will they disperse?** Crowds have a life cycle that includes how they disperse (see *Chapter 1*). Commanders must consider this. It is essential that there be one or more avenues that individuals can use to disperse. These dispersal routes should be clearly marked, visible, and open-ended. There may be a need for US forces to assist the crowd with dispersal. For example, in Los Angeles a 70,000-man march went from east to central Los Angeles. Once at their destination, they engaged in peaceful activities and then dispersed. The LEAs provided transportation back to the assembly area (AA); otherwise, the protesters would have returned to the AA in a disorganized and unsupervised manner.
- **Are there plans for subsequent gatherings?** A crowd may disperse for a short time so participants can take care of personal needs. This could be days or only a few hours.

LEGAL CONSIDERATIONS

2-40. The legal authority in which a peace operation is conducted defines the parameters of the operation. All commanders, leaders, and soldiers must know the legal authority that regulates their operations.

MANDATES

2-41. There are two types of mandates—United Nations (UN) and non-UN. Mandates are the authority under which an operation is conducted. UN mandates are created from resolutions by the UN Security Council (UNSC) or the UN General Assembly (UNGA) that authorize and define an operation. Non-UN mandates are created from treaties, accords, resolutions, or agreements evolving from other international or regional organizations.

STATUS OF FORCES AGREEMENT AND STATUS OF MISSION AGREEMENT

2-42. A status of forces agreement (SOFA) and a status of mission agreement (SOMA) are key documents that define the legal authority and responsibilities of a force and force personnel participating in an operation. A SOFA or SOMA may be a treaty or memorandum of understanding. It is an agreement negotiated between the UN and the host countries, which details the rights, privileges, immunities, and nature of services to be provided to the force and its personnel and their responsibilities and obligations.

INTERNATIONAL LAW

2-43. Legally, international law affects most stability operations and support operations (SOSO). These laws consist mainly of international agreements, treaties, and customary international law, which include the agreements and laws known as the law of war. International agreements prescribe the rights, duties, powers, and privileges of nations relative to particular undertakings. International agreements will affect US participation in SOSO in the—

- Right of US forces to enter a foreign country.
- Status of US forces while in the foreign country.
- Construction and operation of US bases of operations.
- Aircraft overflights and landing rights.
- Processing of claims for any damage to persons and property of the foreign country.

2-44. These agreements or customary international laws govern all aspects of operations carried out. In the absence of a viable HN government or proper international agreement, many aspects of the operations are controlled by domestic (US) or customary international law.

UNITED STATES LAW

2-45. Operations in peace and conflict must comply with US law, whether as a statute, executive order, regulation, or other directive from a branch or agency of the federal government. The *Uniform Code of Military Justice (UCMJ)* applies to questions of military justice. Various statutes, *Executive Order 12333; Section 3, Code of Federal Regulations (CFR);* and DOD and service regulations govern all intelligence-gathering activities. The staff judge advocate (SJA) must actively advise and participate in all activities from initial planning to redeployment.

HOST NATION LAW

2-46. If a viable HN government exists, whether at the national or the local level, all laws of the HN will apply to US forces in that country unless an international agreement provides otherwise. The types of laws that may inhibit US operations are in the fields of immigration, labor, currency exchange, the procurement of goods and services, customs and taxes, and criminal and civil liabilities. A SOFA usually covers these areas before an operation begins. Therefore, planners must understand the law to assess it and determine if it will adversely affect the operation. Assistance may be available from the local US embassy or the command judge advocate; if not, the command may have to rely on other sources for guidance. If local law hinders the operation, the commander must inform the US embassy and request that it negotiate a solution.

CIVIL AUTHORITIES

2-47. There will be times that US forces will deploy to nations where the infrastructure of civil authorities has survived the conflict and is functional. Local authorities and police are responsible for ensuring the safety of their citizens, including the UN High Commissioner for Refugees (UNHCR). US forces assist in this by providing general military security and by facilitating negotiations. In countries where the civil authority infrastructure has collapsed, the UN mandate may require the coalition of forces to help ensure the safety of citizens until their own civil authorities can be rebuilt.

RULES OF ENGAGEMENT

2-48. The rules of engagement (ROE) are the commander's rules for the use of force (RUF). Commanders are responsible for interpreting, drafting, disseminating, and training on the ROE. Ensure that the SJA assists throughout the ROE development process. This will ensure that the ROE will not improperly constrain actions, but are still consistent with domestic and international law and policies and orders of the chain of command.

2-49. Clearly stated ROE are published before Army forces are committed. The highest military authority, with input from subordinate commanders, will continually evaluate the ROE and modify them as appropriate. When the use of NLW has been authorized in the ROE, the decision to use NLW should be delegated to the lowest possible level, preferably the platoon or squad level. This requires that all personnel, not just leaders, have a clear understanding of the ROE and the commander's intent (see *FM 3-22.40*).

2-50. In peace operations, the use of force is restrained while the diplomatic solution is sought through negotiation or mediation. Accordingly, ROE are more restrictive in peace operations, making them more politically sensitive. In today's world, peace operations are usually conducted by a coalition of forces operating under the purview of the UN charter and customary international law. Therefore, the UN may mandate certain restraints. Using overwhelming force could compromise diplomatic efforts to reach a peaceful settlement. Even a single, relatively small engagement could jeopardize the legitimacy of the peacekeeping force and upset negotiations or mediations. Commanders must beware that in any confrontation ROE decisions made by soldiers can have strategic political implications on current and future operations. Therefore, it is vital that leaders and soldiers hold a common understanding of when, how, and to what degree force is to be used. This requires training.

MULTINATIONAL OPERATIONS

2-51. Multinational operations include military forces from different nations. These operations are usually interagency operations. Conducting operations with foreign military partners, like operations with civilian partners, is uncommon to many soldiers, so a clear understanding of this different environment is necessary.

2-52. The majority of US peace operations will be part of a UN peace operation. Their multinational character merits particular attention because national interests and organizational influence may compete with doctrine and efficiency. Commanders can expect contributing nations to adhere to national policies and priorities, which at times complicate the multinational effort.

2-53. When working within a multinational force and with allies, do the following:

- Train coalition partners and allies on tactics and techniques.
- Remember that teamwork and trust are essential.
- Achieve unity of effort.
- Consider equipment and communications capability and interface.
- Strive to eliminate sources of confusion and misunderstanding.
- Remember that military doctrine varies from country to country.
- Remember that national culture influences the way units operate.
- Remember that language barriers represent a significant challenge.

SCALABLE EFFECTS

2-54. Scalable effects concepts are a measured approach in response to a crowd gathering. By recognizing a use-of-force policy, soldiers must be taught and understand that they use the minimum force necessary. Without the appearance of a graduated response, the gathering crowd may consider actions as excessive, causing a possible escalation of hostilities or violence. Do the following to aid in the scalable effects process:

- Try to persuade the crowd to quietly disperse by talking with leaders.
- Use translators as necessary.
- Let the first approach be by local authorities (the mayor or police).
- Pass out handbills requesting that the crowd return home.
- Use video and still cameras to photograph individuals and events for later use at trials.
- Give warnings before moving to the next level of force.

2-55. Ideally, the force should be positioned out of sight of the crowd. Have the troops move into position with few shouted commands. For maximum effect on the crowd, have them form their formation decisively and professionally. Do the following to aid in the formation process:

- Display force in a graduated manner (for example, a UH 60, hardened vehicles, and soldiers with charged weapons).
- Exploit the psychological effects of show of force.
- Demonstrate force (do not use on unarmed civilians).
- Escalate the PSYOP message.
- Highlight the target pointer.
- Demonstrate sniper precision strike capability.
- Employ riot control measures.
- Use NL munitions.
- Use RCAs (if approved).
- Move through the crowd using riot control formations and movement techniques.
- Use the graduated response matrix (GRM)

2-56. With the increased participation of US forces in contingency operations around the world, it accentuates the need to establish procedures for applying graduated military responses to situations that threaten these missions. Numerous GRMs and similar products exist throughout the military. These products graphically portray available responses in a graduated manner. The intent of these products is to give the on-scene commanders a list of options with which to control or defuse a situation before it gets out of hand. Most threats can be eliminated without loss of life or collateral damage by effectively applying the resources available.

PLAN A GRADUATED RESPONSE MATRIX

2-57. The planning for and development of a GRM begins with the mission analysis portion of the military decision-making process (MDMP). Some missions require US forces to enforce treaties or accords to protect the lives of civilians in uncertain or hostile environments (such as noncombatant evacuation operations [NEO]) or to provide large-scale humanitarian assistance. These missions require some sort of graduated response criteria to maintain order and prevent uncertain environments from becoming hostile. The development of a GRM requires the following seven steps:

- Identify the need for a GRM.
- Establish a team to develop the GRM.
- Develop targets.
- Coordinate staff functions.
- War-game.
- Receive command approval.
- Operate deliberately.

Identify the Need for a Graduated Response Matrix

2-58. Not all missions require a GRM. The decision to use a GRM requires careful consideration. Once it is agreed that a GRM is necessary, it requires guidance from the commander regarding the response options available.

Determining the appropriate responses is based on the facts, assumptions, and limitations identified during mission analysis.

2-59. Planners (staff) must agree on the intent of the GRM. The GRM can be used as a training and rehearsal tool. It provides leaders with the most likely vignettes that can be incorporated into a COA analysis, predeployment training, and rehearsals. The GRM can also be used as a handy reference during situations that require graduated responses.

Establish a Team to Develop the Graduated Response Matrix

2-60. Establish a GRM development team with a broad range of skills, knowledge, and professionalism. The following list of individuals makes up a typical GRM team:

- Provost marshal or senior military police advisor.
- FSE.
- Brigade legal officer.
- PSYOP representative.
- Land information warfare or information operations officer.

2-61. Since the GRM is designed to give commanders graduated options for dealing with both hostile and nonhostile threats to the mission, this team composition allows for target selection, application of the ROE, and attack methods using both NL and lethal means.

Develop Targets

2-62. The FSE, in conjunction with the S2 section, develops targets for both lethal and NL attacks. In the case of stability operations, these targets are usually not the conventional specific point or piece of equipment on the ground. They are more situational than specific. The GRM identifies situations or acts that subordinate elements could face during the mission. The sample GRM in *Figure 2-1* shows three possible situations or acts that on-scene commanders could expect to encounter. From the targeting standpoint, these are groups of more specific targets.

2-63. During mission analysis, the fire support officer (FSO) identifies both lethal and NL assets available to the unit. A tactical PSYOP team (TPT) attached to the unit is an example of a NL attack asset that should not be overlooked. The following are examples of what the FSO should look for:

- RCAs.
- TPTs.
- Electronic warfare assets.
- CA teams.
- Information operations (IO) teams.
- Artillery smoke projectiles.
- Aircraft (AH-64s, OH-58Ds, and AC-130s).
- Mortars.

2-64. The lethal assets described could be used in a NL show of force or demonstration to defuse a situation before it requires lethal force. The critical element of this mission analysis is not to focus solely on lethal assets. In

<table>
<tr><th>Crowd Action</th><th colspan="4">Nonlethal Response</th></tr>
<tr><th></th><th>Command Presence/PSYOP</th><th>Show of Force</th><th>Demonstration of Force</th><th>Riot Control Means (if approved)</th></tr>
<tr><td>Unarmed Civilians</td><td rowspan="3">- Ensure that the on-site commander of the TPT directs the broadcast of the dispersal proclamation and/ or passes handbills.
- Escalate the tone of the dispersal proclamation from information to a warning of force.</td><td rowspan="3">- Display force along with escalating the dispersal proclamation.
- Display force in a graduated manner, such as a helicopter hovering over a crowd or soldiers with charged weapons.
- Exploit the psychological effect of a show of force.</td><td>- Do not demonstrate force toward unarmed civilians.</td><td rowspan="2">- Employ RCAs at the point of penetration.
- Use PSYOP to exploit the psychological effect.
- Move through the crowd using riot control formations and movement techniques.</td></tr>
<tr><td>Armed Civilians (knives, clubs)</td><td rowspan="2">- Display force along with escalating the dispersal proclamation.
- Highlight the target pointer.
- Demonstrate sniper precision strike capabilities.</td></tr>
<tr><td>Armed Crowds/ Military (firearms)</td><td>- Do not use RCAs—they may escalate the situation.</td></tr>
<tr><td colspan="5">Hostile intent/hostile act occurs by armed threat.</td></tr>
<tr><th>Crowd Action</th><th colspan="4">Lethal Response</th></tr>
<tr><th></th><th>Sniper Response</th><th>Small Arms Direct Fire</th><th>Aerial Fires</th><th>Indirect Fires</th></tr>
<tr><td>Unarmed Civilians</td><td rowspan="3">- Ensure that target leaders or troublemakers are targeted.
- Use the minimum response necessary.
- Exploit the psychological effect of an attack.</td><td rowspan="3">- Determine that sniper attack is insufficient.
- Consider demonstrating capabilities.
- Exploit the psychological effect of a lethal response.
- Escalate gradually, starting with a small caliber, single round and work up to a large caliber, automatic.</td><td>- Determine that small arms direct fire is ineffective.
- Use the minimum response necessary.
- Use a minimal precision strike initially; use subsequent fires based on the situation.
- Exploit the psychological impact of each strike.</td><td>- Determine that air assets are unavailable or ineffective.
- Use the minimum response necessary.
- Ensure that the response is directed by the on-site commander.</td></tr>
<tr><td>Armed Civilians (knives, clubs)</td><td colspan="2" rowspan="2">- CAS/indirect fires must be authorized by the MACOM commander.
- Consider requesting permission for use when—
• All lesser means have been ineffective.
• There are physical eyes on target.
• Proximity to civilians has been considered.
• Risk to friendly forces/evacuees outweighs the risk of collateral damage.</td></tr>
<tr><td>Armed Civilians/ Military (firearms)</td></tr>
</table>

Figure 2-1. Sample GRM Card

stability operations, the Army wants to prevent acts of hostility first and then be prepared, if necessary, to apply lethal force.

2-65. Graduated responses can range from command presence through the show of force, a demonstration, the use of RCAs, and other techniques (such as the application of lethal force using snipers, small arms, AC-130s, and indirect fires).

Coordinate Staff Functions

2-66. To coordinate staff functions the rest of the GRM team assembles to complete the escalation sequence for each response. PYSOP and legal representatives are critical attendees during the escalation sequencing process. In the area of psychological operations, the TPT must exploit the effects of all responses.

2-67. The legal officer evaluates each escalation of force option and graduated response to ensure that it is consistent with the ROE. The GRM is designed to recommend applications of force consistent with the ROE, yet not limit the leader or individual soldier's right of self-defense. A GRM should clearly show that if a hostile act occurs, lethal options would be first and foremost.

2-68. In the case of lethal responses, the commander's guidance must again be applied. For example, lethal responses are allowed only in self-defense.

War-Game

2-69. Once the types of escalations for each potential graduated response are determined and annotated, the GRM must then be war-gamed. The staff must walk through each act or situation from the on-scene commander's standpoint.

Receive Command Approval

2-70. Once the GRM has been war-gamed, it must be submitted to the commander for approval. This is the final check to ensure that the GRM team has applied the commander's guidance correctly and met his intent.

Operate Deliberately

2-71. Operating deliberately affords commanders ample opportunity to plan and prepare for possible civil disorder situations. Through the effective gathering of information and a working cooperation with local government and police officials, commanders can often be made aware of dates, times, locations, and what groups may assemble before the operation. The purpose is to gather as much information as possible who is involved, where they are assembling, what incident promoted the activity, and what seems to be the prevailing attitude of the assembling crowd.

2-72. Civil disturbance operations are dynamic, ever-changing environments requiring effective communications both up and down the chains of command. They require advance preparation and planning using the established troop-leading procedures (TLP) outlined in *FM 3-19.4*. The eight-step process of TLP aids commanders and their subordinate leaders in planning and preparing for an operation. Most steps can be accomplished concurrently, but rarely is there enough time to go through each step in detail. Despite this, leaders must follow these steps to ensure that nothing is left out of the planning process. As outlined in *FM 3-19.4*, the following are the eight steps of TLP:

- Receive and analyze the mission.
- Issue a warning order.
- Make a tentative plan.
- Initiate movement.
- Conduct a reconnaissance.
- Complete the plan.
- Issue the order.
- Supervise, refine, and rehearse.

CONDUCT HASTY OPERATIONS

2-73. Conducting hasty operations seems to be the norm rather than the exception. Unlike deliberate operations, hasty civil disorder operations are reactionary in nature with little or no time for planning. In these situations, the event of a crowd gathering is already underway. There is very little, if any, advanced warning of the unfolding situation, and commanders are usually put in the position of sending their soldiers into an already volatile, and perhaps hostile, environment.

2-74. Commanders should be in immediate and constant communication with local civil and police authorities. The purpose is to gather as much information as possible about who is involved, where they are assembling, what incident promoted the activity, and what seems to be the prevailing attitude of the assembling crowd.

2-75. It is essential that commanders operating in environments where civil disturbance operations are likely ensure that their soldiers receive training at all levels. Commanders and subordinate leaders need to instinctively rely on the same TLP that they would in a deliberate operation. See *FM 3-19.4*.

CONDUCT TACTICAL OPTIONS

2-76. In the peacekeeping and peace enforcement environment, the use of conventional firearms or the threat of their use may not be the solution to a situation where US forces must separate two belligerent, hostile ethnic groups or prevent a similar group from entering an area that is off limits to them.

2-77. Crowd control options are often combined. Commanders choose their options based on an evaluation of the particular crowd. Commanders select any combination of control techniques and force options they think will influence the particular situation (METT-TC). Commanders must always try to choose the response that can be expected to reduce the intensity of the situation.

Monitoring

2-78. Monitoring an assembled crowd consists of gathering necessary intelligence and watching them to determine progress and development. It is this gathered information that helps commanders select the appropriate actions. Gathered information can often be used to help defuse a situation through persuasion.

2-79. Monitoring is continuous. Without timely information, responses could cause the situation to escalate beyond what it should have or cause a response to be inadequate based on the situation.

2-80. Commanders may task teams with specific missions to monitor crowd activity and note any new developments. Monitoring done by these observation teams gives the commander up-to-the-minute information so he can gauge crowd activity and intent in relation to the overall situation.

2-81. Observation teams monitor crowd activities to gather information. They observe and report on crowd size, location, and mood and on the developing situation. An observation team may consist of a marksman, a radio operator, and an observer equipped with binoculars. They may be posted strategically on rooftops and other high terrain that overlooks the crowd. Sometimes observers use helicopter-mounted observation devices. This also affords security for the operational forces. Commanders must know where their observation teams are located so they are not mistaken for teams of snipers.

2-82. The timely flow of information may allow the commander to influence the outcome of the situation with simple negotiations. Monitoring is appropriate when more decisive action is not feasible because of crowd size or when the intensity of a situation might escalate. It is particularly useful in large, nonviolent demonstrations. Monitoring can serve as an interim measure until more control forces arrive. It includes establishing communication with crowd leaders to convey official interest and intent to the crowd. Monitoring also includes efforts to gain the cooperation of crowd leaders.

2-83. Communication with crowd leaders and participants can help a commander control a situation without more severe measures. If communications exist with crowd leaders, the authorities may be able to divert the leaders or the crowd from their stated or apparent goal. Pressure can be put on leaders to channel the crowd into an area that minimizes disruption to the community and aids crowd control operations. March routes and demonstration areas can be limited to those that will help contain the crowd and reduce their potential for disrupting the community. Pressure can be positive (offering concessions) or negative (with deterrents).

2-84. If the commander can gain the cooperation of crowd leaders, it can decrease the potential for disorder, and if crowd leaders seek cooperation from authorities, officials should try to be accommodating. Crowd leaders can be placed in liaison positions between the crowd and the control force. Crowd leaders can be made responsible for managing the crowd by policing their own activities.

2-85. Taking still pictures or videotaping the faces of individuals within a crowd may prevent or reduce unlawful and violent acts. This could lessen their sense of anonymity. If needed, photographs or videotapes can be used as evidence for prosecution. To be effective, crowd members must see their presence being recorded. The photographer or cameraman should be in uniform to let the crowd know who he or she is. Recorders must be close enough to the crowd to be seen easily, but not close enough to be endangered.

Blocking

2-86. It is not uncommon for protests and demonstrations against the US or its forces to occur outside the main entrances to military installations or US Embassies. Because of the sensitive nature of equipment and information that can be found at these locations, blocking unauthorized entry is critical.

2-87. Blocking is the physical denial of advance upon a facility or an area by a crowd. Commanders may have to task their forces to block a crowd. Blocking physically denies crowd advance. Crowd control formations, particularly line formations, along with barricades can be used to block advancing crowds. Barricades of vehicles, concertina wire, and water-filled barrels are used to block or channel the movement of the crowd.

Dispersing

2-88. Dispersing is taking deliberate actions to fragment an assembled crowd in order to prevent the destruction of property or prevent injury. It is extremely effective against smaller crowds in congested urban environments. However, this may increase and spread lawless activity rather than reduce it. Thus, the commander's forces must control dispersal routes and the areas in which the dispersal will occur. Forces must protect the facilities that could be considered likely targets of opportunity for dispersing individuals or groups. Dispersal of the crowd may require apprehension of small groups still active in the area. Proclamations, shows of force, crowd control formations, and RCAs can disperse crowds.

2-89. Avoid (if possible) the dispersion of crowds into wide-open areas because it gives the crowd the opportunity to grow rapidly in size. The direction of crowd movement is difficult to anticipate and channel.

2-90. Once the crowd has started to disperse, it may occasionally be necessary for the forces to advance, keeping the crowd on the move and in the right direction. A crowd on the move should not be hurried to avoid panic. At no time should the crowd be cornered in a position where there is the perception of no escape. This invokes the "fight-or-flight" syndrome, possibly escalating violent activity.

2-91. Issuing a proclamation (see *Figure 2-2*, page 2-20) can help disperse a crowd. Proclamations officially establish the illegal nature of crowd actions and put the populace on notice that the situation demands special measures. Proclamations prepare the people for the presence of military authority and it tends to inspire respect. Proclamations support law-abiding elements and psychologically bolster military forces trying to restore order. They also demonstrate the gravity of the situation to all concerned and are excellent ways to make a commander's intentions known to a crowd. It is also a good way to reduce crowd size before direct action is taken. Commanders can make a verbal proclamation similar to the following statement: "*In the name of the President of the United States, I command that you disperse and retire peaceably to your homes.*" Such a proclamation may even make direct action unnecessary.

I. DISPERSAL

Unlawful Gathering

"Attention! Attention! This area must be cleared at once! Further unlawful behavior will not be tolerated. Clear this area at once or the necessary force to do so will be used."*

Demonstration

"Disperse and retire peaceably! Disperse and retire peaceably! Attention all demonstrators!"

"The demonstration in which you are participating ends at ______. The permit that was agreed to by the leaders of the demonstration expires at that time. All demonstrators must depart from the ______ by _____. All persons who wish to leave voluntarily may board the buses. These buses will go to the _____. Those who wish to take buses should move to ______. Those demonstrators who do not leave voluntarily by ______ will be arrested and taken to a federal detention center. All demonstrators are urged to abide by the permit."*

Warning of Looting

"Return to your homes! Someone may be looting them at this moment! During a disturbance, criminal activity is at its peak. Your family or your property may be in danger."*

II. EMPLOYMENT OF TROOPS

Employment

"Attention! Attention! Troops are present in this area. They are preparing to advance. Order must and will be maintained. Disperse peaceably and leave this area. To avoid possible injury, leave at once."

"Disperse now and avoid possible injury! Disperse now and avoid possible injury!"*

(Repeat until troops are committed.)

When Troops Are Committed

"Troops are advancing now. They will not stop until this crowd is dispersed and order is restored. To avoid injuries, leave the area at once. Return to your homes as peaceful citizens. Troops have their orders, and they will not stop until the crowd is dispersed. Do not get hurt. Leave this area."*

III. PRESENCE OF CHILDREN

(Used in conjunction with other announcements.)

"Attention! Attention! Do not attempt to cause further disorder. Disperse now in an orderly manner and avoid possible injury to children. Return at once to your homes."*

IV. ADDITIONAL INSTRUCTIONS

*Indicate the method, streets, and direction that the crowd should use when dispersing.

Figure 2-2. Sample Proclamation

2-92. In making a proclamation, a commander may consider imposing a time limit. However, the situation may change, and not imposing a time limit leaves the commander free to choose other actions when he wants. A proclamation must be specific in its instruction.

2-93. If a time limit is stated, it must be a reasonable length of time for the crowd to comply with the instructions. When drafting a proclamation, the commander must consult closely with the SJA. He must use the simplest language possible to maximize the effects of the proclamation. If proclamations must be translated to a local language, the translation must be made with great care.

2-94. The commander at the scene may direct that a proclamation be issued over public address (PA) systems. The force of the words used in the proclamation must be gauged to the composition of the crowd. If the crowd consists of normally law-abiding citizens who are presently assembled to show disagreement with an existing situation, the proclamation requires less force. On the other hand, if the crowd consists of militant rioters, the proclamation requires more force. The text may take a number of forms, depending on the situation.

2-95. A show of force is often a useful measure for dispersing a crowd. When troops arrive, the psychological impact of their arrival can be used. Soldiers can dismount from the helicopters, buses, or trucks in plain sight of the crowd, but they must be far enough away to prevent a provoked attack of thrown objects. The first echelon to dismount from the vehicle secures the surrounding area.

2-96. When small groups are scattered throughout a large disturbance area, a show of force can be made by marching troops conducting motor marches, conducting saturation patrolling, and setting up static posts. Sometimes marching well-equipped, highly-disciplined soldiers in view of a crowd may be all that is needed to convince them to disperse and retire peaceably. On the other hand, a show of force may attract people to an event. It may also provoke a nonviolent crowd into a violent confrontation. Intelligence can help best in preparing a show-of-force response.

Containing

2-97. Containment is the process of limiting a crowd to the area they are presently occupying. It is a suitable option when the crowd must be prevented from spreading to surrounding areas and communities. Additionally, it prevents those outside the already assembled crowd from joining the gathering. Containment is useful when apprehensions become necessary for preventing those in the crowd from escaping. Crowd control formations, perimeter patrols, and barriers are effective methods to accomplish containment. In all instances, caution must be used to avoid the "fight-or-flight" syndrome common to people feeling trapped with no escape.

2-98. Armored vehicles, armored security vehicles (ASVs), and up-armored high-mobility multipurpose wheeled vehicles (HMMWVs) are adaptable to roadblock operations as they can serve as barriers. They also provide added protection for the soldiers inside and those outside manning the roadblocks because they provide an easily accessible barrier for them to crouch behind.

2-99. Military vehicles traveling at close intervals in a column formation next to a crowd are largely a psychological barrier (see *Chapter 6*). They can be used to contain a large, fast-moving crowd. The moving cordon creates a temporary obstacle between the crowd and the line beyond which they will not be allowed to cross. A well-trained driver in a mobile cordon can do a better job than dismounted soldiers. In order to execute safe, quick reversals of direction for mobile cordons on narrow roads, the soldiers executing this formation must be highly-trained vehicle operators.

2-100. By-the-number commands issued over vehicle radios are the most common method for coordinating a cordon movement. Vehicles equipped with PA systems can also prove effective to issue both commands to troops and directives to the crowd. The cordon speed is no more than 5 miles per hour with an interval of at least 20 feet. Blue and red lights, sirens, blinkers, and horns may or may not be used. Armored vehicles can also serve as mobile command posts. When used properly, vehicles provide security, communications, and mobility.

DECIDE TO CONTAIN OR DISPERSE

2-101. Current crowd control doctrine places emphasis on crowd dispersal. Forced dispersal may result in a crowd breaking up into multiple groups that scatter over a large area. This may pose even greater public order problems and may pose a continued threat to control forces. A crowd is often better controlled by means of containment (confining its activities to a given area). A crowd has a limited duration, and its numbers are likely to diminish as individual needs take precedence over those of the crowd.

2-102. During the G8 summit protest in Ottawa, Canada, in June 2002, several thousand people showed up to demonstrate against the evils of world corporate leaders. Canadian Police determined that their best response to this event was a "soft response." Throughout the two-day event, hundreds of officers in standard patrol uniforms operated in four-officer teams, prestaged throughout the city. As the demonstrators began to assemble and eventually took to marching, officers provided escorts both in the front and rear of the enormous crowd along with officers marching to the flanks of the demonstration. This "soft approach" frustrated demonstrators, especially those who had it within their agenda to cause property damage and personal injury. Even more frustrating to the demonstrators was the heavy downpour of rain they endured on both days. Their organizations quickly disbanded out of concern for their own creature comforts. In the end, 10,000 people participated in the protest, with only three arrests made in two days.

USE PATROLS

2-103. Alert, aggressive patrolling of the disturbance area deters the gathering of crowds. The use of saturation patrolling is most effective for this purpose. It allows information to be collected, and it creates the psychological impression of the control force being everywhere at once.

2-104. Standard military transport vehicles provide mobility and communications for area coverage. Soldiers must be deployed with enough vehicles to provide the flexibility to handle all situations in the disturbance area. The vehicles can be modified with sandbags, wire screens, or similar materials to protect against sniper fire and thrown objects. If sniper fire is expected or has been encountered, then up-armored HMMWVs or light armored vehicles should be used for patrolling.

2-105. Armored vehicles patrolling an area of violence provide an added psychological effect and allow troops to maneuver close to snipers to make an apprehension. They may also be used as rescue vehicles to extract hostages or people surrounded by a hostile crowd.

2-106. Varying patrol routes and times for mounted and dismounted patrols keep lawbreakers from being able to pick a safe place or time to act. Patrols are particularly useful in preventing overt arson and violence. Patrol members can also spot and promptly report fires. Whenever possible, military patrols are integrated with civil police patrols. Joint patrols conserve military forces and aid civilian military communications. They also help troops become familiar with an area quickly. In addition, the civil police are available to make apprehensions, if needed. Patrol members must practice proper standards of conduct and fair treatment of civilians at all times. They are performing an important community relation and control function.

2-107. Motorized patrols, because of their speed and mobility, provide timely reconnaissance and broad area coverage. Motorized patrols are in radio contact with TF headquarters (HQ). They can make periodic contact with foot patrols and stationary posts, while patrolling their own areas. Moreover, motorized patrols can respond quickly to calls for help from other patrols and guard posts. Motorized patrols should have at least three vehicles with three men in each. Strength in numbers is necessary for protection. When planning patrol routes, avoid areas where the patrols can be isolated or placed in jeopardy. In addition, motorized patrols are equipped with fire extinguishers to put out small fires, thus reducing the burden on the fire department.

2-108. Besides foot and motorized patrols, air patrols provide a third dimension in directing the overall control effort in the disturbance area. They can perform reconnaissance and surveillance and provide near real time formation over the disturbance area. They are an excellent means of providing timely information to the commander concerning the demonstration area and its surrounding perimeter. They can monitor the actions of rioters, the extent of damage, the status of access routes, locations and conditions of road barriers, and other important conditions.

USE DELAYING TACTICS

2-109. AOs that units are assigned to patrol are often larger than the unit has resources to manage effectively. In addition, units must manage taskings, contingency plans, and other commitments, which will quickly drain available manpower. This problem will manifest itself when factions threaten to riot or gather in large numbers to demonstrate. In order for the commander to stall for time while he shifts unit assets, he may want to delay the arrival of buses and vehicles full of demonstrators. This technique is based on the assumption that demonstrators are moving to the demonstration site by vehicle. Key and secondary routes into the demonstration site must be analyzed by the S2 when he does the IPB.

NOTE: There should be an overwatch when this technique is attempted.

2-110. As demonstrator buses and/or vehicles move toward the demonstration site, soldiers in two HMMWVs block the road just ahead of the approaching vehicles. Once the demonstrators dismount their vehicles and begin to close in on the soldiers, the soldiers leapfrog (see *Figure 2-3*) 1 kilometer farther down the road to repeat the process again. This is repeated as often as necessary. This process, if repeated enough times, should discourage the demonstrators enough that they give up and go home. It should also provide enough time to shift assets to the demonstration site.

2-111. Demonstrators are just as capable of blocking secondary and key routes leading into a demonstration or riot site to prevent relief forces from assisting with riot control as US forces are capable of preventing demonstrators access to the site. One must remember that demonstrators are often well organized and have handheld radios and other items. A great example of this was the 1999 demonstration against the WTO in Seattle, Washington. Police had erected temporary portable fences in the downtown areas to keep demonstrators from disrupting the delegates. Determined to be disruptive, the demonstrators relocated the temporary fences by not allowing police to enter the area. Police were forced to back patrol cars over their own fences to gain access.

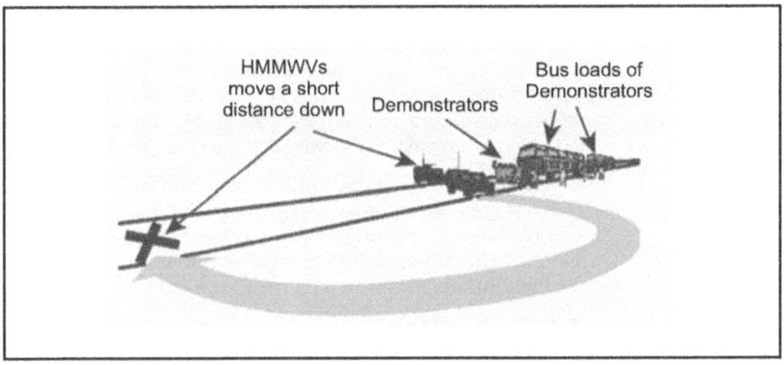

Figure 2-3. Delaying Tactics

Chapter 3

Apprehension, Search, and Detention

During the course of a civil disturbance, some members of the crowd may take part in unlawful activities, such as looting and assault. This will require the civil authorities or attached elements to apprehend, search, and detain people who are participating in the demonstration. Circumstances of the unfolding situation may require US military forces to be called upon to search, take custody of, and detain people who are participating in some form of protest and violating the law. Search, apprehension, and detention operations are conducted to halt these violations and to deter future violations.

LEGAL CONSIDERATIONS

3-1. All apprehensions are made by the civil police force, unless it is not possible for them to do so. Individuals must be known to have committed an illegal act, or there must be probable cause to believe that an individual committed such an act to be apprehended. If it becomes necessary for a military control force to apprehend or temporarily detain such violators, control force members often do so with the approval of the civil authorities. This ensures that all searches, apprehensions, and detentions (if necessary) are conducted within the parameters of the law.

3-2. Commanders operating overseas, similar to Bosnia and Kosovo, may find themselves without an established government or LEAs to work with. Commanders on the ground are then thrust into positions of authority, making them responsible for the actions, conduct, searches, apprehensions, and detentions performed by their troops.

3-3. If US military forces are called upon by the local authorities, certain policies have to be observed during search, apprehension, and detention operations. First, treat all people firmly but with reasonable courtesy and dignity. Remember that participating in a legal demonstration to express views is a right of all people, not just Americans in the US. The action, attitude, and behavior of US soldiers performing these operations are very important.

3-4. Our forces should never be seen as a military policing force on our own soil or as an occupying force in another country. Treating people with contempt, hostility, or excessive force increases the likelihood of resistance and violence. Searching people, placing them under apprehension, and detaining them without probable and just cause or without concern for their constitutional rights create problems and hinder due process. Numerous cases exist where authorities violated an individual's rights, resulting in the prosecution of police authorities or civil suits being waged against the officers and their leaders. This includes the military personnel assisting those officials.

3-5. It is very important that civil law enforcement (if available) be present and supervise all activities. Military personnel conducting a search or making an apprehension must carry out all the procedures carefully within the parameters of their training and the authority given to them. For each search and apprehension performed, the apprehension element must document the specific conduct of the person that violated the law.

3-6. Apprehending officers and supporting US servicemen must provide careful descriptive data for each subject. This description must be sufficient so that, at a later time, the suspected person can be clearly identified as the subject in question. Names and addresses of witnesses to the violation must be obtained. This information can be recorded on *Department of Defense (DD) Form 2708 (Receipt for Inmate or Detained Person).* Apprehension teams must retain and tag each item of physical evidence, such as weapons or stolen goods, that supports the apprehension on *DA Form 4137 (Evidence/Property Custody Document).* The team must give the owner a receipt for all property collected as evidence and retained.

TYPES OF SUBJECTS

3-7. There are two types of subjects (cooperative and uncooperative); they are as follows:

- **Cooperative.** Cooperative subjects actively and willingly comply with instructions. Cooperative subjects are usually compliant and offer no resistance to verbal directions or physical actions. However, if they provide any resistance to authority, it is usually passive. Some examples of this resistance are as follows:
  - Yelling.
  - Screaming.
  - Tightening the body.
  - Pulling away.
  - Letting the body go limp.
- **Uncooperative.** Uncooperative subjects will display preattack cues. Uncooperative subjects will actively resist being apprehended. Often, this resistance comes in the form of physical actions, such as—
  - Assaulting verbally.
  - Refusing to be handcuffed.
  - Refusing to be searched.
  - Attempting to push away, forcing a pursuit.
  - Wrestling with the apprehension team.
  - Striking out at the apprehension team with fists, hands and elbows, and feet.
  - Striking out at the apprehension team with weapons, such as knives, clubs, bats, bricks, rocks, or firearms.

APPREHENSION

3-8. US military forces will be in a supporting role to the local civilian authorities or acting as the control force in civil disturbance operations. When making apprehensions, it is best to use apprehension teams. These teams provide quick, organized responses to developing situations. Teams can be organized at the team, squad, or platoon level depending on the number of apprehensions expected. The intelligence information gathered helps to prepare for events where apprehensions may be necessary.

3-9. Each team (if possible) consists of a civilian police or apprehension officer, security element, and recorder. Team apprehension officers make the actual apprehension with the help of the security elements. If there is not a civilian police officer available to act as the apprehension officer, then it may be an officer or noncommissioned officer (NCO).

3-10. Ensuring that the apprehension process is properly conducted and documented, determine who is to be apprehended. Then inform the subject of why he is under apprehension and supervise his handling.

3-11. Security elements help the apprehension officer handle subjects. Security elements move, restrain, and search subjects under the supervision of the apprehension officer. The primary responsibility of the security elements is to watch the crowd and act as a blocking force, keeping the crowd from interfering with the apprehension. Recorders document the event, preferably by video, and complete *DD Form 2708*. If possible, recorders also photograph the apprehension officer with each subject. This aids in the identification process because it links the apprehension officer to the subject and provides the apprehension officer with documentation of the apprehension for use in court.

3-12. Electronically capturing the events as they unfold is essential in all civil disturbance operations. If personnel and equipment are available, videotape the events as they happen or, at a minimum, take still photographs to provide pictorial documentation that may be used later in court. Recording the scene before, during, and after the apprehension provides necessary evidence and can also help eliminate additional hostile and illegal activities. Having control force members take videos or pictures can help control crowd reaction. It reduces the feelings of anonymity that a crowd often enjoys and makes crowd members less prone to unruly or illegal acts.

3-13. Verbal commands given to the subject should be in a voice and manner appropriate for the purpose and the situation. Commands should be simple, concise, and in an understood language. Verbal commands should be convincing and convey an expectation that the instructions will be obeyed. With desperate, frightened people, sympathy combined with an attitude of firm assurance on the soldier's part may avoid additional violent behavior or trouble. An attitude of forceful authority may be needed with an uncooperative person. Commands given by soldiers must be done in a manner and context so that the subject has no confusion as to what the soldier is ordering them to do.

3-14. A subject will need to be transported from the area. In CONUS operations, civil authorities will usually be responsible for providing the transportation of a subject. Exceptions to this may be if the demonstration occurs on an installation. OCONUS operations may require US forces to provide vehicles, such as trucks, buses, or sedans, that will be used to transport a subject. Whenever possible, vehicles used to transport a subject should be modified with barriers to separate them from the drivers. If a barrier cannot be provided, a guard is to be placed in the vehicle. Passenger compartments are checked for items that could be used as weapons. When guards are required, the guard always sits with his weapon away from the prisoner. Prisoners must be placed where they can be best controlled, but they are not fastened to the vehicle with hand irons or Flex-Cufs. When seat belts are available they must be fastened. See *Soldier Training Publications (STPs) 19-95C1-SM and 19-95C24-SM-TG.*

3-15. Female escorts should always be used to transport a female subject. If female escorts are not available, measures must be taken to avoid false charges of molestation or sexual misconduct. Net control stations should be notified of the departure and arrival times and vehicle mileage readings (before and after). At least two male escorts must be present at all times when a female escort is not available. These procedures also apply when the male and female roles are reversed.

FLEX-CUFS

3-16. The two main positions used to place Flex-Cufs on a subject are the standing-supported and the prone.

3-17. When using the standing-supported position, the search man—

- Escorts the subject to a wall or other sturdy vertical surface (see *Figure 3-1*).

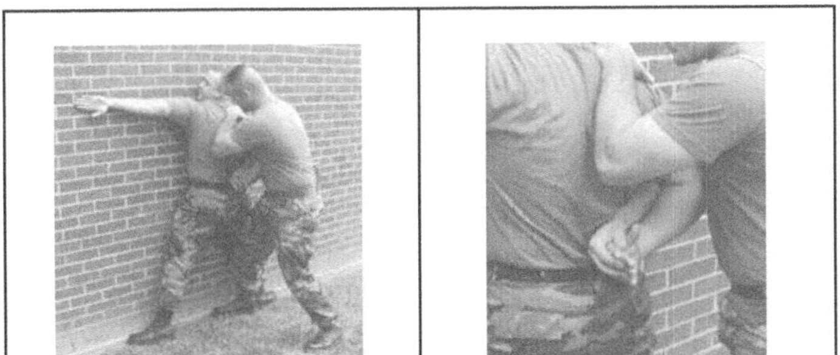

Figure 3-1. Standing-Supported Position

- Instructs the subject (verbally) to do the following:
  - Turn his head away.
  - Place his free arm (palm out, thumb down, and elbow locked) against the surface of the wall.

  - Spread his feet until instructed to stop.
  - Place his heels against the wall.
- Uses the leg closest to the controlled arm to pin the subject's leg (above the knee) to the wall.
- Reinforces verbal commands with pain compliance techniques if the subject fails to follow instructions.
- Maintains a double 90° hold on the controlled arm with his inside hand. With his free hand, the search man grasps the middle portion of the Flex-Cuf, keeping one cuff up and one cuff down.
- Places the first cuff (bottom cuff) on the subject's controlled arm, maintaining wrist manipulation (see *Figure 3-2*).
- Reaches through the free cuff with his outside hand and instructs the subject to place his free hand in the center of his back. With his outside hand, the search man takes the subject's free hand and completes the restraint by securing the cuff. He ensures that there is at least one-finger distance between the cuff and the subject's wrist (see *Figure 3-3,* page 3-6).

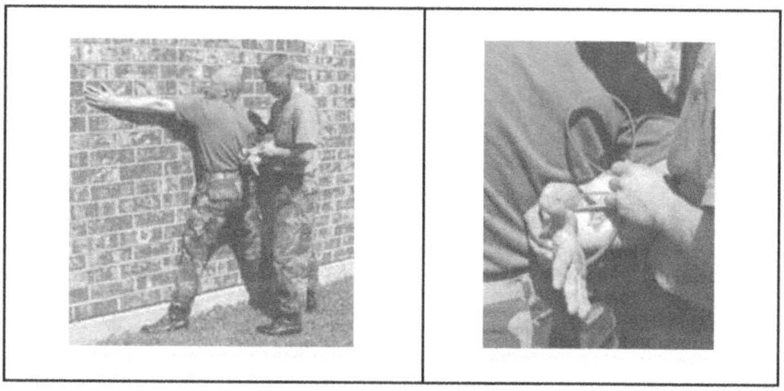

Figure 3-2. Cuffing the First Hand From the Standing Supported Position

3-18. When using the prone position, the search man—

- Grasps the Flex-Cuf with his outside hand, maintains tension on the subject's arm, and places the first cuff on the subject's controlled wrist (see *Figure 3-4,* page 3-6). Next, the search man secures the cuff, keeping at least a one-finger distance between the cuff and the wrist.
- Reaches through the free cuff with his outside hand and instructs the subject to place his free hand in the center of his back (see *Figure 3-5,* page 3-7).
- Reinforces verbal commands with pain compliance techniques if the subject fails to follow instructions. The search man bends the subject's controlled arm (using the natural break of the elbow as a hinge) toward the hand in the middle of the subject's back. With the outside hand, the search man takes the subject's free hand and completes the restraint by securing the cuff.

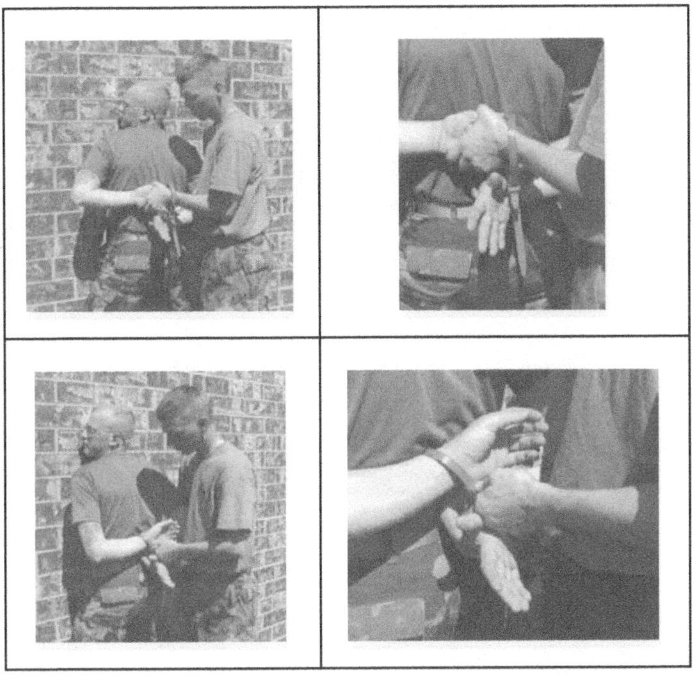

Figure 3-3. Cuffing the Second Hand From the Standing Supported Position

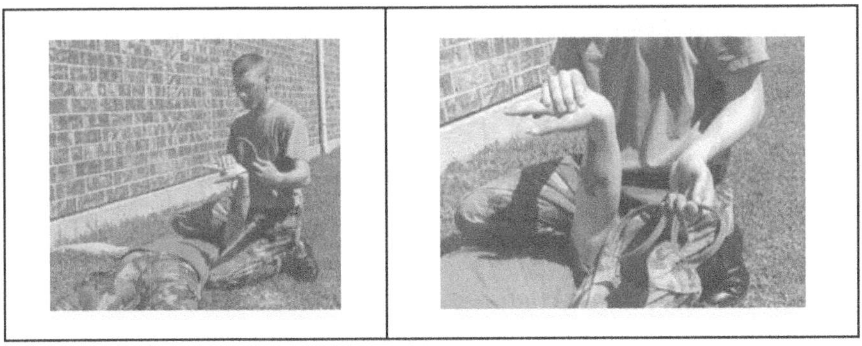

Figure 3-4. Cuffing the First Hand From the Prone Position

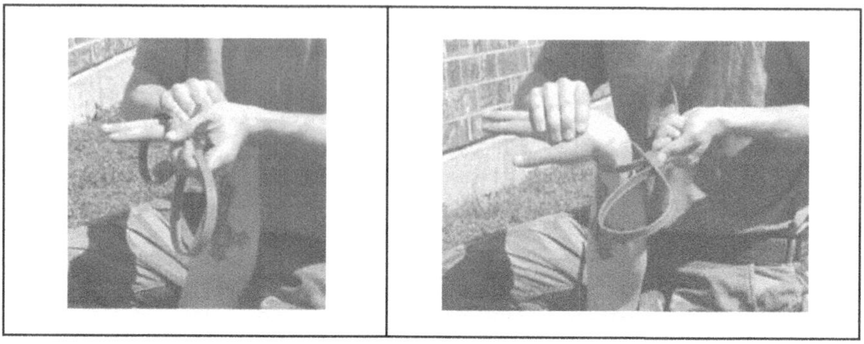

Figure 3-4. Cuffing the First Hand From the Prone Position (continued)

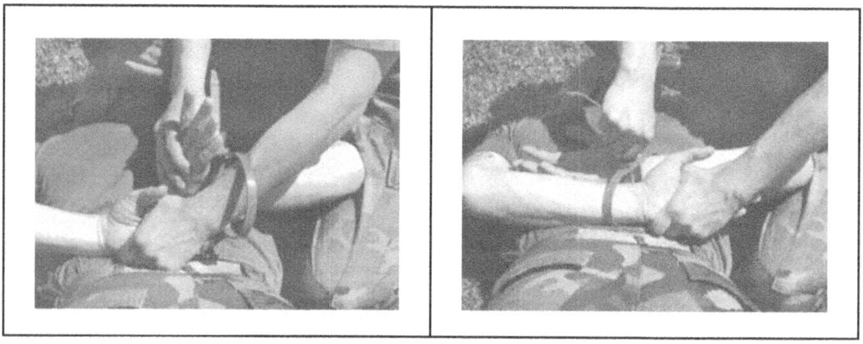

Figure 3-5. Cuffing the Second Hand From the Prone Position

SEARCHES

3-19. Initial contact with the subject is the most dangerous time during any apprehension; therefore, initial contact must be a swift and decisive team effort. Do not give the subject time to think or react. Distract his attention when possible. Apprehension teams are comprised of a search man, threat cover man, and security man (if necessary). Their responsibilities are as follows:

- The search man—
  - Issues verbal commands.
  - Conducts searches.
  - Handles radio communication.
  - Holsters his weapon before approaching a subject and informs the threat cover man that he is holstering his weapon.
  - Performs control procedures and handcuffs the subject.

    - Moves the subject to a safe area.
    - Removes weapons (if any) and sensitive evidence from the subject.
    - Disengages after the subject is restrained.
    - Controls and escorts the subject to a detention area.
- The threat cover man—
    - Provides cover for the search man.
    - Discourages escape and resistance.
    - Discourages hostile interference.
    - Alerts the search man of threats.
    - Resists distractions and maintains target acquisition or scanning.
    - Assesses threats constantly and prepares for the unexpected.
- The security man (if necessary)—
    - Provides cover for the search man.
    - Discourages hostile interference.
    - Alerts the search man of threats.
    - Resists distractions and maintains target acquisition or scanning.
    - Assesses threats constantly and prepares for the unexpected.

Search Principles

3-20. A subject should always be put at a disadvantage and searched immediately, but circumstances may dictate postponing the search until a safer place can be found. At a minimum, a quick pat down can be done en route to a safer location. The longer the apprehension takes, the greater the risk of drawing a hostile crowd and violent acts toward the control force.

3-21. Crowds often sympathize with a subject and try to help him or escalate the activities of the crowd toward violence. A subject's actions may also incite the crowd to violence. The longer a subject is the focal point of an apprehension, the easier it is to stir a crowd to hostile or violent acts. Apprehension teams should choose to move the subject out of the crowd's view. For the safety of the team, always apply Flex-Cufs or handcuffs and escort the subject to an area where a search can be done safely.

3-22. Search team members never operate alone; searches are conducted by a minimum of two team members. Initial steps must be taken to put the subject at a physical disadvantage. Before conducting a search (regardless of rank, sex, position, age, or physical condition), a team member applies the handcuffs or Flex-Cufs. Verbal commands must be short, clear, and distinct.

3-23. A search is conducted on everyone that has been apprehended to ensure that the team and those being apprehended are safe. The search man must abide by the following procedures when conducting a search:

- Search in a clear, secure area (when possible).
- Be ready to counter any resistance or an assault by the subject. Consider the subject armed until proven (by search) otherwise.
- Employ the appropriate level of force immediately, based on the actions of the subject.

- Do not step over the subject when he is at a disadvantage on the ground; go around him and remain out of sight.
- Do a 360° search of the subject's waistline for weapons. Search obvious and inconspicuous areas. Inconspicuous areas such as the groin, armpits, chest, and small of the back must be searched on both men and women; however, do not linger in these areas.

NOTE: Conduct same-gender searches when possible. If mixed gender searches are necessary for speed or security, conduct them in a respectful manner and avoid any action that could be interpreted as sexual in nature. To prevent allegations of sexual misconduct, the apprehension officer shows careful control of the soldiers who perform mixed-gender searches.

- Maintain security.
- Brief subjects quickly on what to do and who to follow.
- Remember that when conducting a search, a subject's body can be divided for a more efficient and thorough search. For example, search from head to toe or from side to side.

NOTE: The search man announces the items found and secures weapons and contraband.

- Military forces (when supporting civil authorities) may perform searches of people and property that are not usually subject to military law during a civil disturbance.
- Remember that it is always better for civilian police to search civilians and civilian property (when possible).
- Remember that if military forces are called upon to support police it is usually a last resort and there are not enough civilian officers available to accomplish the mission.
- Military personnel may search people incident to an apprehension.

3-24. The prescribed methods of searching someone are the frisk, the standing, and the prone. The method of search used largely depends on the situation and the mission. Searches are made not only of the subject and his personal effects such as purses, bags, and wallets, but also of the immediate surrounding area. This prevents the subject being searched from grabbing a weapon or destroying evidence. It may also be necessary for military forces to search private property, including vehicles, if there is—

- Reasonable belief that a person has committed or is committing a violent crime and is hiding in a building or a vehicle.
- Reasonable belief that a vehicle contains weapons or instruments of violence.
- Probable cause for searching a building or a vehicle.
- Probable cause to believe that (unless immediate action is taken) evidence of a crime will be destroyed before a warrant can be obtained.

NOTE: Search teams should always consist of two personnel—one to conduct the search, and the other to provide security and overwatch for the soldier conducting the search. Before being searched, a subject should always be restrained and put at a disadvantage by

securing his hands with handcuffs or Flex-Cufs. Ask the subject if there is anything in his pockets that could hurt the searcher, such as knives or needles.

Frisk Search

3-25. The frisk search (also known as the stop and frisk, pat down, and terry stop) is a search incident to a lawful stop. Subjects are not under apprehension at this point, but if it were an apprehension situation, handcuffs or Flex-Cufs would be applied first. Frisk searches are used when there is reasonable belief that a subject may be armed with anything that could be considered a safety threat.

3-26. Legal and regulatory circumstances for conducting a frisk search are covered in the *Manual for Courts-Martial (MCM)*. When a lawful stop is performed, the person stopped may be frisked for weapons when that person is reasonably believed to be armed and dangerous.

3-27. The primary purpose of a frisk search is safety. Frisk-search techniques are to be used only on a compliant subject. Frisk searches only cover the outer garments. Do not search the pockets of the individual unless a weapon is suspected. A frisk search begins when the search man informs the subject of the intent to search.

- During the preparation phase, the search man—
  - Tells the subject that he must perform a search for weapons for the subject's safety and his own.
  - Tells the subject to turn and face away from him and instructs the subject to keep his hands out to his sides where they are visible.

NOTE: The cover man providing security moves around the subject and positions himself out of reach and in front of the subject.

  - Instructs the subject to spread his feet until instructed to stop (more than shoulder width) and point his toes outward compromising his balance (see *Figure 3-6*).

NOTE: Do not kick at the subject's feet to spread them apart; it is unnecessary and unprofessional.

  - Tells the subject to bring both of his hands back to the small of his back with palms out, thumbs up, and the back of his hands together.
  - Instructs the subject not to move (see *Figure 3-7*).
- During the performance phase the search man—
  - Approaches the subject from behind (if possible) at a 45° angle, off center. The search man approaches with caution to ensure that his feet are to the outside of the subject's stance. See *Figure 3-8*, page 3-12.
  - Stays just inside arms reach and grips the subject's fingers with his support hand. The search man firmly grips at least one finger on each of the subject's hands (both thumbs, both index fingers, or an index and middle finger). An overhand or an underhand grip can be used effectively.

Figure 3-6. Approaching the Subject to Perform a Frisk Search

Figure 3-7. Giving Instructions to Subject Before Frisk Search

NOTE: The primary purpose of gripping the fingers is to control the subject's hands while performing the frisk search (see *Figure 3-9*, page 3-12). If the subject is going to resist or become aggressive, he will attempt to free his hands first. If the subject attempts to pull his fingers from the search man's grip, the search man must respond instantly by increasing his grip. If the grip on the fingers is lost, the search man should immediately push away from the subject and create a safe distance. This will allow the search team to respond with the necessary force to regain control of the situation.

Figure 3-8. Performing a Frisk Search

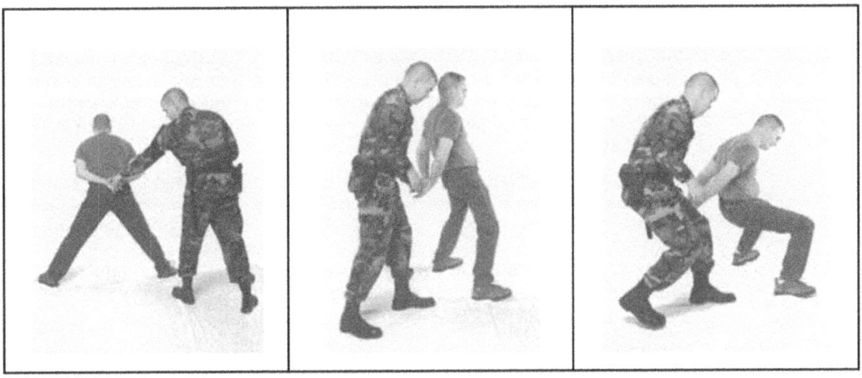

Figure 3-9. Gripping the Subject's Fingers During a Frisk Search

- Ensures that the subject always looks away from him.
- Places his strong hand on the outside of the subject's elbow (similar to the handcuffed escort position).

NOTE: The search man searches one side (the front, back, or side) at a time from head to foot.

- Divides the subject (visually) straight down his middle and does not cross the line while conducting the search.

NOTE: The search man is just feeling or searching for possible weapons, so an open hand should be used.

- Works at arm's reach. It is the sense of touch not the sense of sight that is being used. While performing the search, search men should stay behind and to the side of the subject. This provides the search man with a tactical advantage and prevention from being struck by the subject's elbows or knees if he should begin to resist.

3-28. When searching below the subject's knees, the search man takes a half step back and to the outside. He squats down, but does not drop his knees to the ground.

3-29. If no weapons are found and the subject is to be detained, the hands are secured with Flex-Cufs and the subject is escorted to the holding area. If the subject is to be released, collect information (name and address) and escort him away from the crowd.

Standing Search

3-30. Safety is the primary concern when dealing with a subject under apprehension. Therefore, the search man must protect himself by restraining subjects under apprehension before conducting the standing search. All subjects under apprehension have the propensity to turn violent or hostile. Putting a subject at a physical disadvantage provides safety for the subject and the search man. It limits the subject's ability to hide, destroy, or remove any evidence he may have on him.

3-31. Apply Flex-Cufs to the subject quickly during the apprehension. It is impossible to predict what a subject will do when put under apprehension. A subject may initially be cooperative and compliant, but as the subject becomes aware of his situation he may be overcome with emotion and begin to struggle or pull away.

3-32. There is no legal or regulatory prohibition against opposite-gender searches. Common sense prevails and whenever possible, searches should be done by soldiers of the same gender to avoid obvious questions and complaints. Once a subject is put at a physical disadvantage by cuffing and is more easily controlled, it affords the necessary time to get someone of the same gender there to conduct the search.

- During the preparation phase, the search man—

NOTE: Standing searches always begin with the subject in the handcuffed escort position (see *Figure 3-10,* page 3-14).

  - Takes his outside hand (farthest from subject) and places it on the outside of the subject's elbow while his inside hand (closest to subject) has a reverse-handshake grip of the subject's hand. At this point, both hands are on the same arm (see *Figure 3-10,* page 3-14).
  - Ensures that his position is on the same side of the subject's body being controlled (behind and to the side). When changing sides, change hand position.
  - Instructs the subject to spread his feet until instructed to stop (usually more than shoulder width) and point his toes outward, compromising the subject's balance.

Figure 3-10. Handcuffed Escort Position

NOTE: Do not kick at a subject's feet to spread them; it is unnecessary and unprofessional. If the subject becomes noncompliant or aggressive, pain compliance is usually the best tool to use, but only use enough force to gain compliance.

  - Uses his outside hand to conduct the search. His inside hand is always used to maintain control during the search.
- During the performance phase, the search man—

NOTE: Only one side (the front, back, and side) of the subject is searched at a time from head to toe (see *Figure 3-11*).

  - Divides the subject (visually) (straight down the middle using the spine as the centerline). He does not cross this line while conducting the search. Searches must be systematic, from head to foot. If there is an interruption, it must be started all over again.
  - Searches with an open hand, using a rub down or squeezing method. Do not only pat down like in the frisk search.
  - Works at arms reach. It is the sense of touch, not the sense of sight that is being used. To prevent the possibility of being struck by an elbow or knee, the search man stays behind and to the side of the subject.
  - Takes a half step back and to the outside when searching below the subject's knees. The search man squats down, but does not drop his knees to the ground. Maintaining positive control is essential throughout the search, so never the reverse handshake is never released.
  - Change sides when satisfied that all items have been removed from the side in which the search started. Without losing positive control, the search man conducts the search on the other side of the subject.

Figure 3-11. Performing the Standing Search

NOTE: Changing sides is accomplished by moving the outside hand (the searching hand) to the subject's uncontrolled handcuffed hand located on the outside. The outside hand and forearm will cross over the top of the reverse-handshake grip of the inside hand (see *Figure 3-12,* page 16).

- Crosses over the reverse-handshake grip with the outside hand and grips the subject's uncontrolled handcuffed hand in a reverse-handshake grip.
- Releases (quickly) the first reverse-handshake grip moving that hand to the subject's elbow as he sidesteps to the side yet to be searched. The search man repeats the standing search technique used on the first side of the subject.
- Gives strong verbal commands and uses only the force necessary to control the subject if he resists or becomes aggressive at any time during the search.
- Removes all items found during the search to be evaluated as potential evidence or a weapon. Items not considered a weapon or of evidentiary value will be returned to the subject at a later time.

Prone Search

3-33. Safety is the primary concern when dealing with a person under apprehension. Therefore, the search man must protect himself by restraining subjects under apprehension before conducting this search. Remember that all subjects under apprehension have the propensity to turn violent or hostile. That is the reason a prone search would be necessary. After performing an arm bar takedown on a noncompliant subject, it is especially important for the search man to put him at a physical disadvantage to provide for the subject's safety and his own. By doing so, the subject's ability to hide, destroy, or remove any evidence is limited.

Figure 3-12. Performing the Standing Search (Changing Sides)

3-34. The safest way to handcuff or Flex-Cuf a noncompliant subject is in the prone position, when the search man has greater control of the subject. It is usually the safest position in which to conduct a search

3-35. Handcuff the subject quickly during the apprehension. It is impossible to predict what a subject will do when put under apprehension. Often a subject may initially be cooperative and compliant; however, as the he becomes aware of his situation, he may become overcome with emotion and begin to struggle or pull away.

3-36. There is no legal or regulatory prohibition against opposite-gender searches. Common sense prevails and whenever possible, searches should be done by the same gender to avoid obvious questions and complaints. Once a subject is at a physical disadvantage by cuffing and is more easily controlled, it affords the necessary time to get someone of the same gender there to conduct the search.

NOTE: The subject is put into a physical disadvantage when he is taken facedown to the ground using an arm bar takedown and cuffed.

- During the preparation phase, the search man—
  - Ensures that the Flex-Cufs are not too tight.
  - Rotates his entire body around the subject's shoulder and center at the outside of the subject's hip with his shin over the subject's shoulder and one knee on the ground. At least one knee should remain over the subject's back (near side), with the other on the ground.
  - Pins the subject down by pressing his knee on the side of the subject's back or down on his closest elbow if resistance or aggression is encountered. He simultaneously grips the subject's closest handcuffed hand, grabs it with a reverse handshake, and applies pain for compliance.

- Uses strong, forceful verbal commands, such as "Stop resisting, stop moving around."
- Ensures that the subject's feet and ankles are crossed.

CAUTION

Stay alert. If the subject uncrosses his feet and ankles, it is usually a sign of preattack.

- Removes all items found during the search so they can be evaluated as potential evidence. Items not considered a weapon or of evidentiary value will be returned to the subject at a later time.

- During the performance phase, the search man—

NOTE: The search man must be systematic from head to foot while conducting the prone search. If interrupted during the search, it is important to start from the beginning to ensure that nothing is missed. It is better to go over the same area two or more times than to miss something that can be used as a weapon.

NOTE: The search man searches only one side (the front, back, or side) of the subject at a time until that side is complete.

- Divides the subject (visually) straight down the middle, using the spine as the centerline.

NOTE: He does not cross this line while conducting the search.

- Searches with an open hand, using a rub down or squeezing method for best results. He does not just pat down the subject like in the frisk search.
- Works at an arm's reach, searching only the area of the subject's body that is not in contact with the ground. The sense of touch, not the sense of sight, is being used.
- Stops the search as the subject's knees are reached. He taps one of the subject's legs and firmly commands him to raise his foot up and back, allowing for an inspection of the bottom of his foot. He then searches from the knee to the ankle area. This is performed on both the subject's legs and feet.

NOTE: At this point, the entire back side of has been searched. What remains to be searched is the front side of the subject (the chest and front of the hips, waist, groin, and thighs).

- Pivots back and faces the subject. Using the hand closest to the subject's head, the search man reaches across the subject's back and firmly grips the shoulder, rolling the subject up on the his side toward him to expose the areas not yet searched.
- Rolls the subject back down on his chest once all areas have been searched.

3-37. The search man must be completely satisfied that he has found everything. He must be especially sensitive to areas where weapons are often concealed like the waistband, the small of the back, ankles, pockets, shoes, the tongue of the shoes, and the groin area.

NOTE: Before removing subjects from the area, a brief summary must accompany them. The summary must include the subject's name, what he is suspected of doing, the time and date of the incident, and who saw him do the action.

DETENTION

3-38. Civil authorities must provide adequate detention facilities for all subjects. Authorities must be prepared to detain large numbers of people. They may choose to expand existing detention facilities or set up temporary facilities to accommodate the extra load. If possible, large-scale arrests are delayed until sufficient detention facilities have been set up.

3-39. If US military forces are committed to support local authorities, commanders should coordinate with them to ensure that adequate detention facilities are available and to learn their locations and capacities. If there are more detainees than civil detention facilities can handle, civil authorities may ask the military to provide support by setting up and operating temporary facilities. Army correctional facilities cannot be used to detain civilians. A temporary military detention facility can be set up if—

- Federal troops have been employed according to provisions.
- The TF commander has verified that available civilian detention facilities can no longer accommodate the number of prisoners awaiting arraignment and trial by civil courts.
- The Army Chief of Staff has granted prior approval.

3-40. Use of the temporary facility ends as soon as civil authorities can take custody of the detainees. Military forces are responsible for the custody, health, comfort, and sustenance of all detainees in their facilities until custody is transferred to civil authorities. Temporary facilities cannot be used to confine people arraigned or convicted in civil courts.

3-41. Women are detained in their own facility. Temporary facilities are supervised and controlled by officers and NCOs trained and experienced in military correctional operations. Females guard females. If two female guards are unavailable, use two male guards or one male and one female guard (see *FM 3-19.40)*.

3-42. The same operational procedures that apply to the management of an installation confinement facility (see *FM 3-19.40*) apply to the management of temporary detention facilities, except for training, employment, and administrative discipline. Guards and support personnel under direct supervision and control of military officers and NCOs need not be trained or experienced in military correction operations. They must be specifically instructed and closely supervised in the proper use of force, custodial procedures, and the completion of military and civilian forms and reports that may be used. Soldiers that may be tasked to operate a temporary detention facility should be familiar with the forms and reports used for civilian apprehensions too.

3-43. Temporary facilities are set up on the nearest military installation or on suitable property under federal control. Ideally, the facility should be close enough to the disturbance area to minimize transportation and escort needs. However, it should be far enough away not to endanger those being detained. Whenever possible, existing structures are adapted for this use, but construction may be needed to provide segregation for ensuring effective control and administration.

3-44. The basic structure must include the following:

- Search areas.
- Holding areas for incoming men and women.
- A processing area.
- Holding areas for men and women who have committed misdemeanors.
- Holding areas for men and women who have committed felonies or are violent.
- A holding area for detainees' property and evidence obtained during a search.
- A holding area for administrative support and records.
- A medical station.
- Latrines.

For more information on these structures see *FM 3-19.40.*

3-45. Facility personnel must ensure that proper sanitation is maintained. When large numbers of people are detained or processed through a facility, sanitation becomes a problem. Medical personnel must conduct regular health inspections to detect unsanitary practices and conditions.

3-46. Facilities must be organized for a smooth flow of traffic. Processing stations must be set up so there is a linear or circular sequence of movement. These stations are set up out of sight of the holding areas. If possible, they are separated by a door to reduce noise. To reduce distractions, each station may be partitioned. Detainees may be more cooperative if they are out of sight of each other.

3-47. Injured people are given prompt medical treatment and transportation to medical facilities when necessary. A medical aid station for screening detainees and treating minor injuries is set up inside or next to the detention facility. Treatment areas, however, must be out of sight of the processing and holding areas. If possible, access to the medical facility bypasses the holding and processing areas. Facility personnel may also consider setting up a separate holding area for injured detainees.

3-48. Detention facility operation plans must contain emergency procedures. As a minimum, the following areas must be addressed:

- Fire evacuation.
- Disturbance control.
- Facility security.

NOTE: Depending on the situation, other emergency procedures may be needed.

3-49. Upon arrival at the detention facility, the detainee is logged in and searched, regardless of whether a complete search has already been conducted by the apprehending team in the field. Separate search areas are set up for men and women. Weapons, contraband, flame-producing devices, suspected evidence, money, and high-value items are confiscated. Medications are also confiscated. Medical personnel screen detainees that have had medications confiscated. Receipts are provided for any property or evidence that is confiscated. Confiscated items are tagged, and the items are stored in a controlled-property area.

3-50. As a detainee is brought to the facility, a file is initiated. Assigned detainee case numbers are used on all paperwork, such as logs, evidence tags, reports, and visual documentation. All paperwork (including photographs) that began outside the detention facility is marked with the case number. Facility personnel may also use hospital identification tags. Using indelible ink, the case number is written and attached to the tag on the detainee's wrist. Different colors may be used to identify different subject classifications, such as misdemeanors, felonies, or violent offenders. If opposing factions are involved, procedures are established to ensure that members of opposing factions are not processed together or detained in the same holding area.

3-51. After a detainee has been searched and classified, he is taken to a processing station where his paperwork is processed. If subjects passively resist by going limp, they may be moved by wheelchair. This reduces the number of escorts needed and the fatigue to facility personnel. Paperwork is reviewed to ensure that all information is complete, including charges, witnesses, and reasons for the apprehension.

3-52. Equipment such as height charts, scales, fingerprinting equipment, and cameras must be available for completing the police report and identifying the subject. Detainees may use aliases and not carry identification, so pay special attention to obtaining as much information as possible about the detainee's physical characteristics.

3-53. After processing is complete, the detainee is placed in the appropriate holding area. Paperwork for the detainee is forwarded to the administrative section. Files are reviewed for completeness and to determine the disposition of the detainee. Information from the file may be placed in a computer to find repeat subjects. Computers may be used to file criminal information only.

3-54. Custody transfers and release procedures must be coordinated with civil authorities and the appropriate legal counsel. Every effort must be made to arraign subjects quickly. The purpose of a detention facility is not to keep people off the streets, but to aid in processing subjects through the legal system. To speed up the release process, planners should consider issuing citations or subpoenas for minor offenses.

3-55. Military force members may have to respond to writs of habeas corpus. Writs are court orders addressed to a prisoner's custodian. Such a writ directs the custodian to bring the prisoner to court to determine the legality of the prisoner's apprehension and detention. Custodians must ensure that all documentation concerning the case is brought to court with the detainee.

3-56. Military personnel must obey writs issued by federal courts. For writs issued by a state court, the custodian or his legal advisor should respectfully reply that US authority is holding the prisoner. The Office of the SJA can answer any questions and explain the correct procedures.

Chapter 4

Riot Shield and Riot Baton Techniques

A soldier may be required to be part of a riot control formation where his primary weapon is a riot baton and his only protection a riot shield. A riot baton in the hands of a well-trained soldier is an invaluable weapon in a crowd control situation and may also be the most appropriate weapon. A riot baton is a very versatile weapon that can be used as an offensive and/or defensive weapon. The riot shield offers protection from the top of the head to just below the knees.

OVERVIEW

4-1. Use of the riot baton and riot shield is based on METT-TC and the commander's choice-of-force option. Formations committed to a riot with riot batons must have lethal overwatch, lethal weapons, and NLW employed within the formation and reserve forces with additional NL and lethal capabilities. These forces are positioned to rapidly reinforce the control force formation.

4-2. During the termination phase of the disturbance (when the violence has subsided and is expected to remain that way), soldiers are placed in a defensive posture. They may be equipped with a riot shield and a riot baton to perform their tasks. The riot shield is primarily a defensive weapon, which may be used in an offensive mode. It is primarily used as an offensive weapon when the formation is in close contact with rioters. The riot baton can also be used when formations are on the offensive and in contact with the crowd. However, crowd control elements employed with riot shields and riot batons must have lethal weapons and NLW available.

WARNING

Riot control formations are never employed with riot shields and riot batons only. There must always be lethal overwatch elements in position and lethal and NL capabilities within the riot control formation with a ready reserve force armed with lethal and NL capabilities.

RIOT SHIELD

4-3. In the hands of a properly trained soldier, the riot shield has proven to be the first line of defense for a formation. The riot shield may be used as an offensive weapon when in contact with an aggressive crowd.

4-4. The riot shield may be held with one arm. It is secured to the weak side arm with a self-sticking strap system and a handle. This system is designed to make the riot shield an extension of the soldier's arm. The riot shield is held at a slightly inward angle. This allows for debris thrown at the soldier to fall harmlessly to the ground (see *Figure 4-1*).

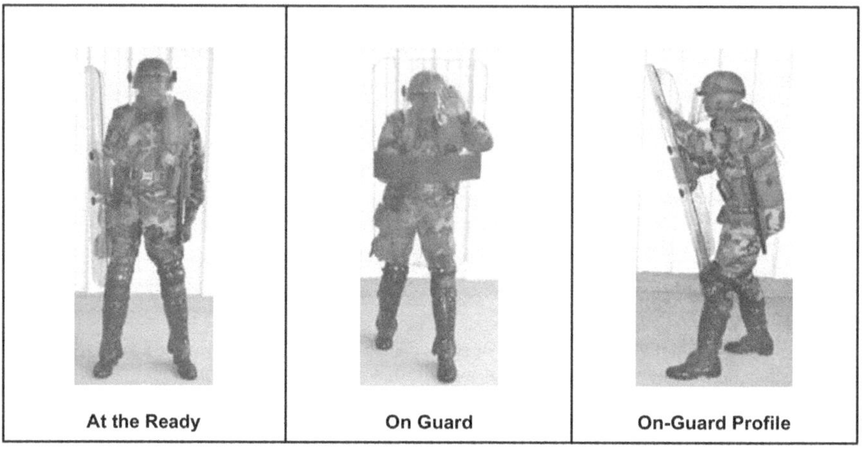

Figure 4-1. Riot Shield Positions

RIOT SHIELD CARRIES

4-5. The two basic riot shield carries are as follows:

- **At the ready.** The at-the-ready position is used while in a column formation or on the move toward hostilities. This position is designed for rapid movement. It should not be used when in contact with a crowd.
- **On guard.** The on-guard position is used when the riot control formation is in contact with the crowd or contact is eminent. It is used primarily in the following formations:
  - Line.
  - Wedge.
  - Echelon.
  - Diamond.
  - Circle.

DEFENSIVE AND RETENTION TECHNIQUES

4-6. The primary purpose of the riot shield is for defense. However, each riot shield holder must be proficient in riot shield retention techniques.

DEFENSE

4-7. The first line of defense for the formation is the riot shield. It is designed to withstand strikes from objects thrown by the crowd. In addition to protecting the riot shield bearer, it also protects other members of the formation from thrown objects.

RETENTION

4-8. When the control force is in contact with the crowd, the crowd will attempt to strip the riot shield from the bearer. When a subject grabs the top of the riot shield (see *Figure 4-2*), the riot shield holder slaps the riot shield with his strong hand near the subject's hands and gives the following commands (while continuing to slap the riot shield until the subject releases it):

- "Get back."
- "Get away."
- "Stop."

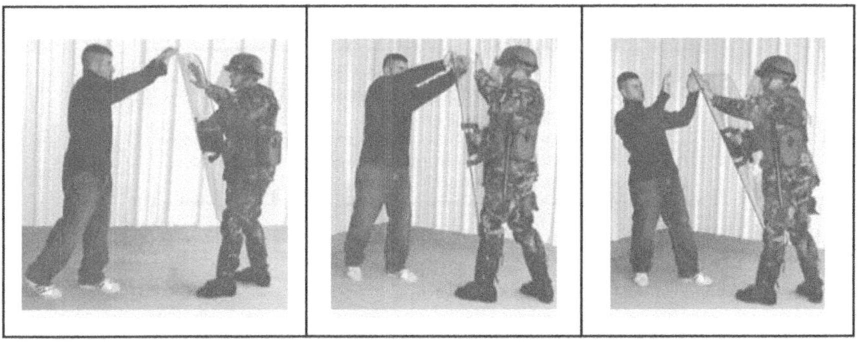

Figure 4-2. Riot Shield Retention Technique (Top Attempt)

4-9. If the riot shield is grabbed from the bottom (see *Figure 4-3,* page 4-4), then the riot shield holder forcefully drops to one knee, pinning the subject's fingers and/or hands to the ground. This must be done quickly and with force.

RIOT BATON

4-10. A riot baton in the hands of a properly trained soldier is a formidable weapon. However, it must be used in conjunction with other measures to be most effective. Soldiers must be trained with the riot baton to the point that its various techniques are automatic to them. This training must also include

Figure 4-3. Riot Shield Retention Technique (Bottom Attempt)

learning the vulnerable points of the human body so they can avoid areas that may cause permanent injury or death when struck.

WARNING

The riot baton is never raised above the head to strike a subject in a club fashion. Not only is it likely to cause permanent injury; but it also gives an unfavorable image of the control force. The soldier is also vulnerable to an attack on his rib cage when his arm is raised.

TARGET AREAS OF THE HUMAN BODY

4-11. The target areas of the human body are divided into three areas that represent the primary-, secondary-, and final-target areas. These areas are color coded as green (shown as gray in *Figure 4-4*), yellow (shown as white in *Figure 4-4*), and red (shown as patterned gray in *Figure 4-4*).

Primary-Target Areas

4-12. The primary-target areas are those areas which, when struck or restrained, cause the least amount of trauma to the body. These include the meaty or muscular areas of the body, such as the—

- Foot, shin, and instep.
- Inside and outside of the thigh.
- Lower abdominal region.
- Fore- and upper-arms.

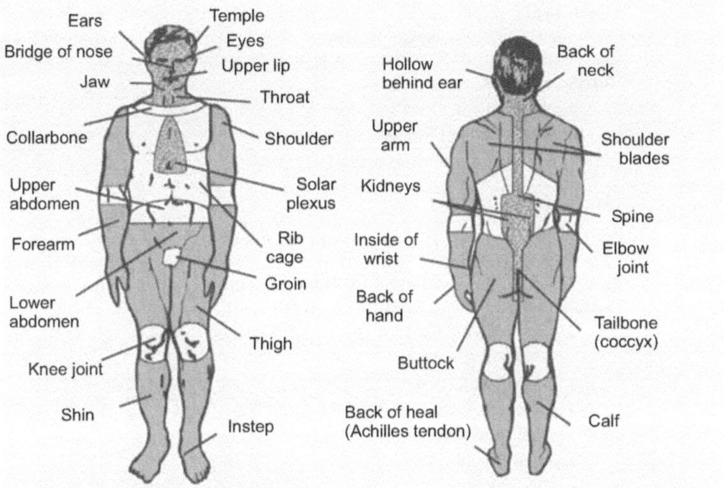

Green (lowest level of trauma) Primary-target areas:

Reasoning: Minimum level of resultant trauma. Injury tends to be temporary rather than permanent; however, exceptions can occur when applying as a restraint technique.

NOTE: When performing a block with a riot baton, the whole body is a green area, except for the head, neck, and spine.

Yellow (moderate level of trauma) Secondary-target areas:

Reasoning: Moderate level of resultant trauma. Injury tends to be more permanent, but may also be temporary when applying as a striking technique.

Red (deadly force) Final-target areas:

Reasoning: Highest level of resultant trauma. Injury tends to be permanent rather than temporary and may include unconsciousness, serious bodily injury, shock, or death.

Figure 4-4. Escalation of Trauma Chart

Secondary-Target Areas

4-13. The secondary-target areas are those areas of the body which, when struck or restrained, cause a moderate level of trauma to the body. This type of trauma tends to be longer lasting because the time to heal is longer than with other parts of the body. If these areas are struck with a riot baton, serious injury may occur. However, some of these areas are used for control or restraining holds that tend to lessen the threat of injury. Some secondary targets are the—

- Collar bone and shoulder blades.
- Elbows and knees.
- Upper abdominal region.

Final-Target Areas

4-14. The final-target areas are those areas which, when struck with a riot baton, cause serious trauma that tends to be fatal or critical. However, there are some areas that (when used for restraining only), will not cause death or serious injury. Some final-target areas are the—

- Front and back of the head (eyes, ears, nose, upper lip, and hollow behind the ear).
- Neck and throat.
- Upper chest.
- Xiphoid process.
- Spinal column.
- Groin area.

JOINTS OF THE HUMAN BODY

4-15. The joints of the human body are often described as hinges and, when manipulated in a certain fashion, allow for better control of the subject being restrained, especially if he is aggressive. When applied correctly, minimal pressure is needed to control the subject. When a subject attempts to resist, he creates additional pain for himself. By anatomically locking out these hinges, the subject will be forced to comply. When restraining a subject, emphasis must be placed on locking out two of the three joints of the upper body or two of the three joints of the lower body. The joints are as follows:

- **Shoulder.** Apply pressure to the shoulder to prevent the subject from freely moving the upper portion of his body.
- **Elbow.** Apply pressure on the elbow or just above the elbow on the ulner nerve.
- **Wrist.** Apply pressure by pancaking the hand to twist or lock out the wrist. Apply pressure at the break of the wrist and push the subject's thumb toward a stationary object (the wall or the ground) to lock the hand out.
- **Hip.** Apply pressure to the hip to knock the subject off balance or alter his center of gravity.

- **Knee.** Apply pressure to the knee by pushing outward on it to cause the subject to be placed off balance, or use the knee as a hinge for pain compliance.
- **Ankle.** Apply pressure to the ankle by clasping and turning it in a counterclockwise manner to lock the hinge out.

TYPES OF RIOT BATONS

4-16. The Army is currently fielding two types of riot batons, wooden and expandable. The most common riot baton is the 36-inch hickory riot baton with thong. There is also the 24- to 36-inch expandable riot baton, which has been added to the nonlethal capabilities set (NLCS). Regardless of which riot baton the soldiers are equipped with, all techniques for blocking and striking are identical. Similar to the riot shield, the riot baton offers soldiers both offensive and defensive capabilities.

4-17. Soldiers must be properly trained in all blocking and striking techniques. Improper use of the riot baton by an untrained soldier has the potential for creating a greater problem than what already exists.

NOMENCLATURE OF THE RIOT BATON

4-18. The riot baton is divided into four basic parts: the long end, long portion, grip portion, and grip end (see *Figure 4-5*). Although the expandable and wooden riot batons differ, their nomenclature is quite similar.

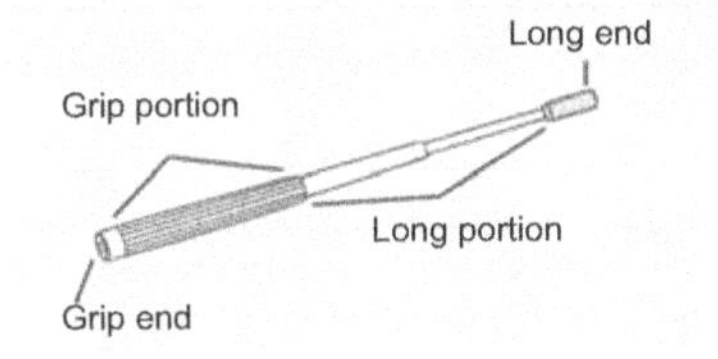

Figure 4-5. The Riot Baton

TYPES OF RIOT BATON CARRIES AND POSITIONS

4-19. As with any other weapon that the soldier will use, there are certain carries and positions from which the riot baton may be used offensively or defensively.

Two-Hand Carry

4-20. The two-hand carry is the most common carry for the riot baton. It is intended for executing all blocks and returning with strikes. When the expandable riot baton is in the collapsed position (at 24 inches), the long end of the riot baton is angled downward approximately 45°. When it is extended to 36 inches the long end is angled upward approximately 45°.

4-21. The wooden riot baton is secured with the soldier's hand by means of a leather thong. He hooks the thumb of his strong hand (palm down) in the thong, pulls the thong across the back of his hand, and grips the riot baton. This provides him with a very secure grip (see *Figure 4-6*).

4-22. The two-hand carry is intended for executing all blocks and returning with strikes. The soldier grips the riot baton with both hands. His strong hand grips the short end of the riot baton, and his weak hand grips the long portion of the riot baton, 2 to 4 inches from the end. His palm is down with his strong hand pulled in close against his strong side hip.

4-23. Wherever the long portion of the riot baton is pointed is where the soldier will strike or jab. The riot baton should be kept at a 45° angle to the ground. All blocks and two-hand jabs and strikes are executed from the two-hand carry.

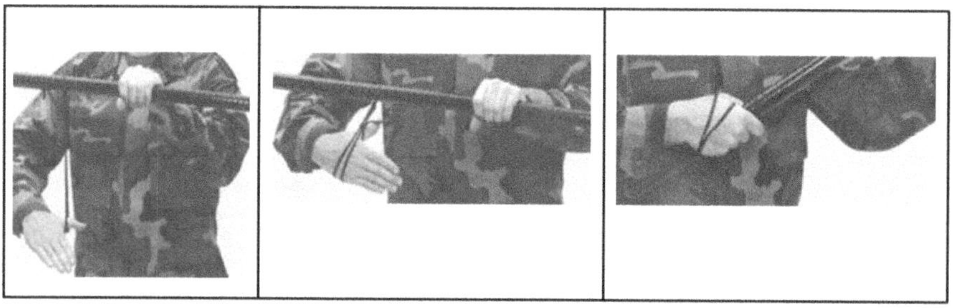

Figure 4-6. Two-Hand Carry

Outside-Arm Carry (Expandable Riot Baton Only)

4-24. The outside-arm carry (see *Figure 4-7*) is used with the expandable riot baton in the collapsed (24-inch) position only. Assuming a wide-based defensive stance (knees slightly bent and feet shoulder width apart), a soldier draws the grip portion of the riot baton with his strong hand. Without striking out or widely swinging the riot baton, he brings the long portion up against the outer part of his strong arm (the tricep area). The soldier's weak hand should be raised in a position to protect and block strikes from an aggressor. It is from this carrying position that a soldier can effectively use both the front and rear strikes. Front and rear strikes should never be performed with an expandable riot baton in the expanded position or with a 36-inch wooden riot baton.

CAUTION

The expandable and 36-inch wooden riot batons provide a greater risk of injury to the soldier's wrist.

Figure 4-7. Outside-Arm Carry

Vertical High-Profile Carry

4-25. Assuming a wide-based defensive stance, the soldier draws the riot baton with his strong hand by the grip portion. Once the riot baton has been drawn from the carrier, the soldier swings the riot baton in a downward motion, fully extending the arm and holding the riot baton down and parallel to the strong side leg. The soldier's weak hand is raised in a position to protect and block strikes from an aggressor (see *Figure 4-8*, page 4-10). Although this carry can be used with both riot baton systems, it is best suited for the expandable riot baton collapsed to 24 inches. The vertical high-profile carry clearly shows the soldier with a drawn riot baton in an effective defensive posture and ready for attack. A drawn riot baton in the hands of a trained soldier may offer just the kind of deterrence needed to stop the activity of an aggressor.

Vertical Low-Profile Carry

4-26. Assuming a wide-based defensive stance, the soldier draws the riot baton with his strong hand by the grip portion. Once the riot baton has been drawn from the carrier, the soldier swings the riot baton in a downward motion, fully extending his arm and holding the riot baton down and parallel to his strong side leg. His weak side hand is raised in a position to protect and block strikes from an aggressor (see *Figure 4-9*, page 4-10). Although this carry can be used with both types of riot batons, it is best suited for the expandable riot baton collapsed to 24 inches. This position portrays the soldier in an effective defensive position ready for an attack, but with a possible hidden weapon. The major difference between the vertical low-profile and high-profile carry is that in the low-profile carry the soldier hides the riot baton behind his strong side leg, allowing for some element of surprise to a potential aggressor. A drawn riot baton in the hands of a trained soldier may offer the deterrence needed to stop the activity of an aggressor.

Figure 4-8. Vertical High-Profile Carry

Figure 4-9. Vertical Low-Profile Carry

RIOT BATON BLOCKING TECHNIQUES

4-27. There are five basic types of riot baton blocking techniques that are executed from the two-hand carry position.

High Block

4-28. The high block is very effective in blocking a downward, vertical strike directed at the top of the head and shoulders. This type of downward, vertical strike may be from a club, pipe, or similar object. The high block is a two-step movement. To perform a high block the soldier performs the following steps:

Step 1. Raises the riot baton straight across his body until it is parallel to the ground (see *Figure 4-10*).

Step 2. Brings the riot baton straight-up in front of his body until the long portion of the riot baton is in front of his forehead (approximately 2 inches from his head). The riot baton should be in a horizontal position above his head and slightly angled down toward his weak hand, as it absorbs the shock from the strike (see *Figure 4-10*). The fingers of his weak hand should be open and behind the long portion of the riot baton when blocking the object to protect his fingers from being smashed by the strike.

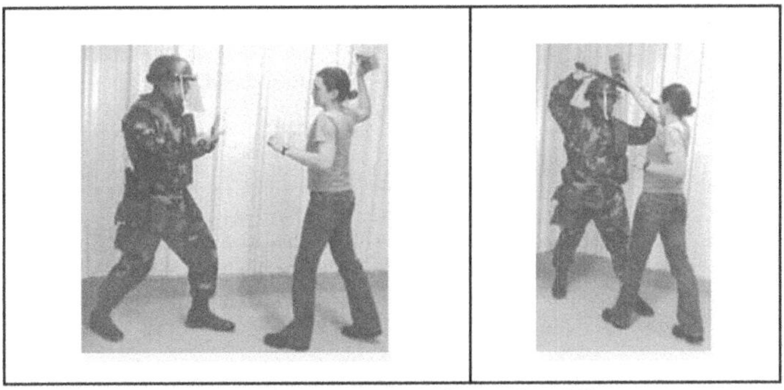

Figure 4-10. High Block (Steps 1 and 2)

Low Block

4-29. A low block is very effective in blocking an upward vertical strike directed at the groin, lower abdomen, chest, or chin. This upward vertical strike may be from an individual's foot, knee, or fist (for example, an upper cut to the chin). To perform a low block, the soldier performs the following steps:

Step 1. Raises the riot baton straight across his body until it is parallel to the ground (see *Figure 4-11*, page 4-12).

Step 2. Brings the riot baton straight down in front of his body, bending at his knees and keeping his body upright until the long portion of the riot baton is just below his knees (see *Figure 4-11*). The riot baton should be horizontal and parallel to the ground as it absorbs the shock from an upward strike. His weak hand fingers should be open and behind the long portion of the riot baton when blocking the object to protect his fingers from being smashed by the strike.

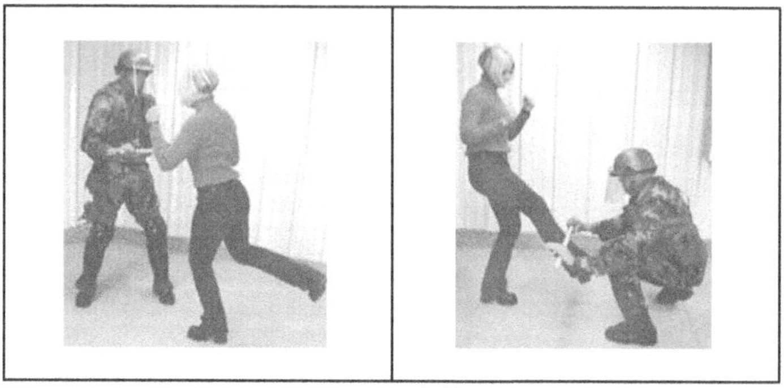

Figure 4-11. Low Block (Steps 1 and 2)

Strong Side Block

4-30. A strong side block is very effective in blocking a horizontal strike directed at the strong side of the head, neck, chest (or flank), or hip area. The horizontal strike may be from an individual's foot, fist, elbow, knee, or weapon. To perform a strong side block, the soldier performs the following steps:

Step 1. Raises the riot baton straight across his body in a vertical position where the grip portion is in his strong hand and the long portion is in his weak hand with the fingers straight up (see *Figure 4-12*).

Step 2. Turns his body laterally by rotating his hips toward his strong side and moving the riot baton across his body, keeping the riot baton vertical (see *Figure 4-12*).

4-31. The strong side block provides protection for a soldier from the front to the rear of his body. A riot baton held in the vertical position absorbs the shock of a strike coming from the strong side. The fingers of his weak hand should be open and behind the long portion of the riot baton, when blocking the object to protect his fingers from being smashed by the strike.

Weak Side Block

4-32. A weak side block is very effective in blocking a horizontal strike directed at the weak side of the head, neck, chest, or hip area. The horizontal strike may be from an individual's foot, fist, elbow, or knee. To perform a weak side block, the soldier performs the following steps:

Step 1. Moves the riot baton in a vertical position toward his weak side, ensuring that the grip portion is pointing straight down and the long portion is pointing straight up.

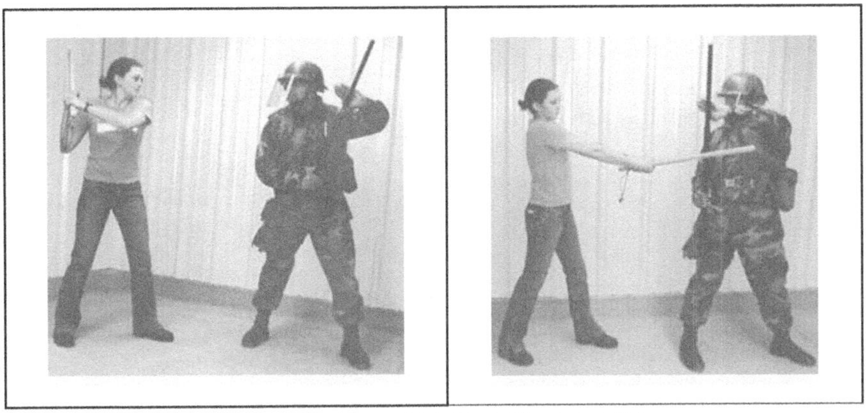

Figure 4-12. Strong Side Block (Steps 1 and 2)

Step 2. Keeps the fingers of his weak hand open and behind the long portion of the riot baton when blocking the object to protect his fingers from being smashed by the strike. At the moment of contact with the opposing strike, the blocking surface of the long portion should be at a right angle to the opposing object.

Step 3. Assess the situation following a weak side block, and take appropriate follow-up action, as needed.

Middle Block

4-33. A middle block is very effective in blocking a front horizontal strike directed at the face, throat, chest, abdomen, or hip. The horizontal strike may be from an individual trying to tackle near the waist or shove or punch at the face. To perform a middle block, the soldier performs the following steps:

Step 1. Moves the riot baton toward the middle of his body at a 45° angle, keeping the long portion pointed up and slightly forward of the grip portion (see *Figure 4-13*, page 4-14).

Step 2. Keeps the fingers of his weak hand open and behind the long portion of the riot baton when blocking the object to protect his fingers from being smashed by the strike. However, his weak hand may have to grip the long portion when pushing an individual away or forcing an individual down on the ground if the individual is trying to tackle.

RIOT BATON STRIKING TECHNIQUES

4-34. Seven strikes can be executed using the riot baton. Two of these strikes come from the outside-arm or vertical carry. The other five strikes come from the two-hand carry.

Figure 4-13. Middle Block

One-Hand Forward Strike

4-35. The one-hand forward strike is a very quick and effective offensive strike. It is usually employed as a countermeasure to an attack from the front that has been blocked effectively.

4-36. Target areas can be as high as the outside upper arm, down to the outer thigh region. An effective one-hand forward strike must be aggressively quick, with the purpose of distracting, disabling, or altering behavior. Therefore, those employing this technique must be sensitive to what part of the riot baton they use to strike the individual. Strike the target with the last 2 to 4 inches of the long end for optimal results.

4-37. A one-hand forward strike may be employed in one of the following three ways. When selecting available targets, use the principles in *Figure 4-4,* page 4-5.

- A horizontal manner.
- A downward diagonal manner to destabilize an individual.
- A vertical manner in front of the body to clear an aggressor's hands.

NOTE: When the one-hand forward strike is used in a vertical manner, keep the long portion straight up, not angled forward. By keeping the long portion up rather than angled forward, the chance of striking the individual in the face or head is significantly reduced. For the counterstrike to be effective, follow through with the technique when the baton makes contact with the aggressor and/or his object to destabilize the him. A pattern of movement such as a

forward shuffle, forward pivot, strong side step, or rear pivot may enhance this technique.

> **WARNING**
> **Do not execute the one-hand forward strike with the wooden or expandable riot baton in the extended position. It has been shown to cause injury to the wrist of the user.**

4-38. When using the vertical or outside-arm carry position (see *Figures 4-7,* page 4-9 and *4-8,* page 4-10), the soldier quickly moves the riot baton across his body using the strength and power of his hips, moving from his strong side to his weak side (see *Figure 4-14,* page 4-16). Ending this strike leaves the riot baton positioned under the weak side armpit in preparation for the one-hand reverse strike (see *Figure 4-14*). When delivering this strong side strike technique (in a horizontal or downward diagonal manner), ensure that the palm is facing up. In a vertical delivery, the palm will be toward the chest rather than up. Pause following the one-hand forward strike and assess the situation. Take appropriate follow-up action, as needed. If the situation does not call for additional strikes, avoid striking again.

One-Hand Reverse Strike

4-39. The starting position for the one-hand reverse strike is under the weak side armpit (see *Figure 4-14*). The one-hand reverse strike is used as a follow-up strike and is quick and effective.

4-40. Target areas can be as high as the outside upper arm, down to the outer thigh region. An effective one-hand reverse strike must be aggressively quick, with the purpose of distracting, disabling, or altering behavior. Therefore, those employing this technique must be sensitive to what part of the riot baton they use to strike the individual. Strike the target with the last 2 to 4 inches of the riot baton for optimal results.

4-41. A soldier may employee a reverse strike in one of the following three ways:

- A horizontal manner.
- A downward diagonal manner to destabilize an individual.
- A vertical manner.

4-42. In front of the body to clear an individual's hands. When the one-hand reverse strike is used in a vertical manner, keep the long portion straight up, not angled forward. By keeping the long portion up rather than angled forward, the chance of striking the individual in the face or head is significantly reduced. When selecting available targets, use the principles in *Figure 4-4,* page 4-5.

4-43. The one-hand reverse strike is considered a follow-up strike to the one-hand forward strike. As a follow-up strike, it is important that soldiers are taught the discipline of assessing the target. An aggressor may not require another strike, based on his condition and actions after enduring the one-hand forward strike.

Figure 4-14. One-Hand Forward and Reverse Strikes

4-44. When using the one-hand reverse strike, the solder quickly moves the straight baton across his body, using the strength and power of his hips (from weak side to strong side). When the strike is completed, the soldier's arm will be back in the outside-arm carry position (see *Figure 4-7,* page 4-9). The soldier ensures that his palm is facing down when delivering the counterstrike in a horizontal or diagonal manner. This will prevent injury to the user's wrist. In a vertical delivery, the palm will be toward the chest rather than up. Following the one-hand reverse strike, assess the situation and take appropriate follow-up action, as needed. If the situation does not call for additional strikes, the soldier should avoid striking again.

4-45. When the one-hand reverse strike is used in a vertical manner, keep the long portion straight up, not angled forward. By keeping the long portion up rather than angled forward, the chance of striking an individual in the face or head is significantly reduced. For the counterstrike to be effective, follow through with the technique when the baton makes contact with the aggressor and/or his object to destabilize him. A pattern of movement such as the forward shuffle, forward pivot, strong side step, or rear pivot may enhance this technique.

WARNING

This strike will not be executed with a 36-inch wooden or expandable riot baton in the extended position. It has been shown to cause injury to the wrist of the user.

Two-Hand Strong Side Horizontal Strike

4-46. Starting from the two-hand carry position, the two-hand strong side horizontal strike can be aggressively quick. The purpose of this strike is to create distance, distract, disable, or alter behavior. Strong side horizontal strikes are quick and effective offensive strikes, usually employed as a countermeasure to close frontal attacks that have been effectively blocked. When selecting available targets, use the principles in *Figure 4-4,* page 4-5.

4-47. The two-hand carry position enables soldiers to use riot batons at the 24- or 36-inch length. To effectively employ this strike, use the strength and power of the hips to thrust the riot baton in a horizontal manner. Simultaneously, pull back with the weak hand as the strong hand drives the grip end toward the target, striking a rib or the abdominal region of the aggressor (see *Figure 4-15,* page 4-18). Following the two-hand strong side horizontal strike, assess the target before following up with another strike. An aggressor may not require another strike, based on his condition and actions after enduring the one-hand forward strike.

NOTE: A pattern of movement such as a forward shuffle or forward pivot may enhance this technique.

Two-Hand Weak Side Horizontal Strike

4-48. An effective two-hand weak side horizontal strike must be aggressively quick to create distance, distract, disable, or alter behavior. The weak side horizontal strike is a quick and effective offensive strike usually employed as a follow-up strike to the strong side horizontal strike. It is a countermeasure designed to close frontal attacks that have been effectively blocked. When selecting available targets, use the principles in *Figure 4-4.* Maintaining a strong defensive stance after employing a strong side horizontal strike puts the soldier in position to execute a two-hand weak side horizontal strike.

4-49. Use of the two-hand carry position enables soldiers to use riot batons at 24 or 36 inches. To effectively employ this strike, use the strength and power of the hips to thrust the riot baton in a horizontal manner. Simultaneously, pull back with the strong hand as the weak hand drives the grip forward toward the aggressor, striking a rib or the abdominal region of the target (see *Figure 4-15*). Following the two-hand weak side horizontal strike, it is important to assess the aggressor. An aggressor may not require another strike, based on his condition and actions after enduring the one-hand forward strike.

NOTE: A pattern of movement such as a forward shuffle or forward pivot may enhance this technique.

Two-Hand Front Jab

4-50. The two-hand front jab starts from the two-hand carry position (see *Figure 4-16,* page 4-18). It is aggressively quick and creates distance, distracts, disables, and/or alters behavior. Front jabs are effective offensive strikes, usually employed as a countermeasure to charging and overpowering frontal attacks. When selecting available targets, use the principles in *Figure 4-4.*

Figure 4-15. Two-Hand, Strong Side, and Weak Side Horizontal Strikes

4-51. Use of the two-hand carry position enables soldiers to use riot batons at 24 or 36 inches. Use the strength and power of the hips to effectively employ this strike in a slightly downward or horizontal manner. Simultaneously, use both arms to shoot the long end of the riot baton straight out from the body to the target area. Immediately, pull back the riot baton to the two-hand carry position (see *Figure 4-16*).

Figure 4-16. Front Jab

4-52. Forward jabs drive the long end of the riot baton toward the aggressor, striking a rib, a hip flexor, or the abdominal region (see *Figure 4-4*, page 4-5). Following the two-hand front jab, it is important that soldiers assess the aggressor before following through with another strike, as the aggressor may not require another strike.

NOTE: A pattern of movement such as a forward shuffle will enhance the power of this technique.

Two-Hand Rear Jab

4-53. Starting from the two-hand carry position, the two-hand rear jab is aggressively quick and creates distance, distracts, disables, or alters behavior. Rear jabs are quick and effective offensive strikes usually employed as a countermeasure to charging and overpowering attacks from the rear. When selecting available targets, use the principles in *Figure 4-4,* page 4-5.

4-54. Use of the two-hand carry position enables soldiers to employ riot batons at 24 or 36 inches. To effectively employ this strike, use the strength and power of the hips to thrust the riot baton in a slightly downward or horizontal manner. Simultaneously, turn the head to the rear to see and assess the target. Use both arms to shoot the riot baton straight back from the body to the targeted area and immediately pull back the riot baton to the two-hand carry position (see *Figure 4-17*).

NOTE: A pattern of movement such as a rear shuffle will enhance the power of this technique.

Figure 4-17. Rear Jab

Two-Hand Middle Strike

4-55. The two-hand middle strike starts from the two-hand carry position. It is aggressively quick and creates distance, distracts, disables, or alters behavior. It is proven to be an effective follow-up technique to a two-hand weak or strong side block or to destabilize an individual by pushing him back and away. The two-hand middle strike is used as a two-count movement. When selecting available targets, use the principles in *Figure 4-4*.

4-56. To perform a two-hand middle strike (see *Figure 4-18*, page 4-20) the soldier does the following:

Step 1. Brings the riot baton up to chest level and slightly off parallel to the ground.

Step 2. Steps forward aggressively, simultaneously thrusting the long portion of the riot baton forward in a horizontal manner. Both arms are fully extended at the end of the movement with a quick snappy return to the step 1 position. Use the entire body to generate the power.

NOTE: It is important that soldiers assess the aggressor before applying a follow-up strike. If the soldier steps forward as he employs the middle strike, it will afford greater impact and power.

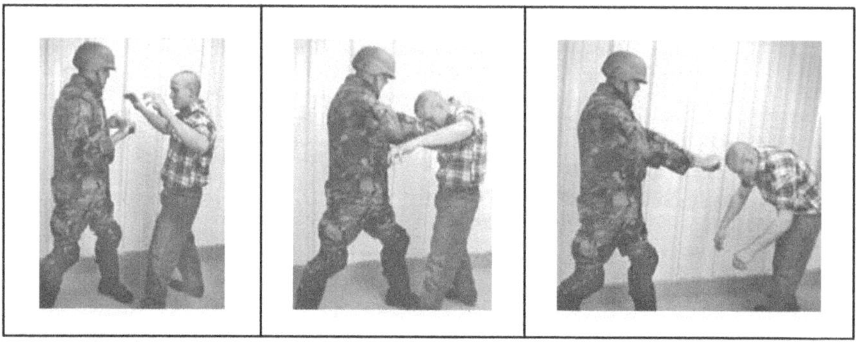

Figure 4-18. Middle Strike

Riot Baton Retention

4-57. A common technique an aggressor may use is grabbing for a soldier's riot baton or, in some cases, even taking the riot baton and using it against the soldier. This causes a disruption or distraction.

4-58. Maintaining physical control of the riot baton is essential for the safety of the soldier and the control force as a whole. Aggressors will generally attempt to grab the riot baton where they can get the strongest hold. Soldiers must resist the instinct of getting into a tug-of-war match over the riot baton, which often proves to be ineffective.

4-59. Using the momentum of the aggressor to pull the riot baton away, the soldier simply steps into or closer to the aggressor with the weak side foot. With the momentum created, the aggressor's hold on the riot baton is lessened. The soldier then rapidly uses the tracing-C technique to recover the riot baton.

4-60. A soldier using the tracing-C technique does the following:

Step 1. Pushes up on the riot baton.

Step 2. Pulls the riot baton around in an upward swing as if tracing a C in the air from bottom to top with the long end.

NOTE: This movement seems to wrap the aggressor's arms around themselves (see *Figure 4-19*).

Step 3. Drives (immediately with a quick and forceful movement) the long end of the riot baton straight down, as if slicing the C in half (see *Figure 4-19*).

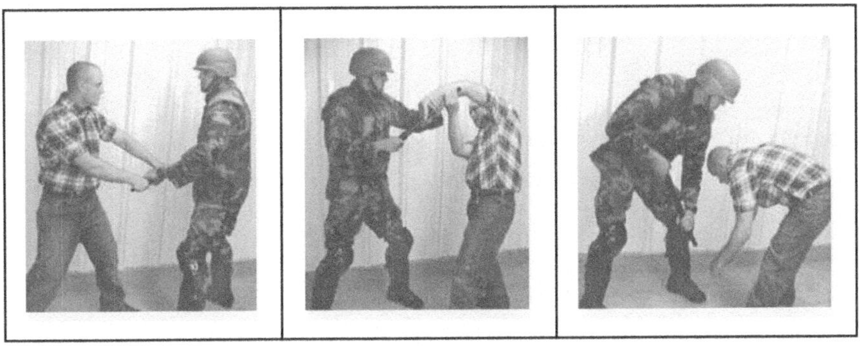

Figure 4-19. Riot Baton Retention

4-61. Once the soldier reaches the top of the C and the aggressor is tied up, it is impossible for the aggressor to maintain hold of the riot baton. Further actions such as a strike may or may not be necessary. Soldiers must be able to assess the situation and react accordingly.

WORKING AS A TEAM

4-62. Soldiers never have to rely solely on their own skills or abilities in riot control operations. Although individual skills were discussed in this chapter, it is important to remember that well-trained soldiers employing offensive, defensive, riot shield, and riot baton techniques will perform the collective task of crowd control well.

4-63. Riot control operations are dynamic and often chaotic. Soldiers armed with riot shields provide a first line of defense while those armed with riot batons and firearms provide overwatch as the formation moves toward engaging a hostile crowd. To work effectively and in concert, soldiers must be properly trained, drilled, and exercised. If an aggressor approaching a line of skirmishers violently grabs a riot shield, help is required to maintain control (see *Figure 4-20*, page 4-22). A soldier armed with a 36-inch wooden riot baton or an expandable riot baton (extended to 36 inches) can be of assistance. The soldier simply moves forward into a gap between the riot shields where he can quickly strike the top or side of the riot shield, slicing the long end across or straight down onto the hands and fingers of the aggressor (see *Figure 4-18*).

4-64. Riot batons are not the only way to effectively knock an aggressor's hands from a riot shield. Soldiers armed with 12-gauge shotguns or M16/M203s can just as effectively smash the hands or fingers of a persistent aggressor with the barrel of their weapons using the technique described above.

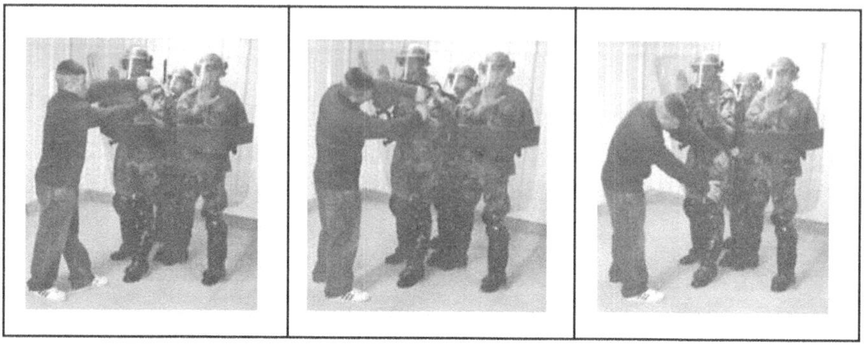

Figure 4-20. Clearing a Riot Shield

NONLETHAL MUNITIONS

4-65. With the inclusion of NL munitions, commanders now have additional tools that afford excellent standoff capabilities. This is critical because commanders now have intermediate options for dealing with a hostile crowd. No longer will a commander issue a proclamation, have it ignored by the crowd, and then be forced to use lethal munitions to backup the proclamation. Now he can use NL munitions to gain the same standoff distance but without the catastrophic results of lethal munitions.

4-66. NL munitions include (but are not limited to) a 12-gauge shotgun and 40-millimeter rounds. These blunt munitions have the capability of effectively engaging targets as close as 5 meters and as far as 50 meters. Range is dependent upon each round. These munitions are used from a distance to dissuade, discourage, and disperse the crowd. If the crowd disperses from 50 meters and soldiers are in close contact with the crowd, then the mission has been successful.

TAP-DOWN TECHNIQUE

4-67. Commanders must be familiar with the characteristics and capabilities of all rounds to use them effectively. When 12-gauge and M203 weapons are used within a line of skirmishers (usually located behind the riot shields), it is imperative that a method is developed, trained, and practiced to effectively operate as a team. Operators of these weapons may encounter problems in riot control situations when trying to identify and effectively engage legitimate targets. The tap-down technique is used to overcome target identification and engagement with NL munitions. If ordered to engage targets or the situation authorizes an engagement, soldiers armed with NL munitions may step forward, level their weapons, and fire upon legitimate targets (see *Figure 4-21*).

4-68. A soldier that is armed with a NLW and is using the tap-down technique must first tap on the shield holder's shoulder. Tapping on the shield holder's shoulder alerts him that the gunner is preparing to fire. In response to the tap, the shield holder drops to one knee while keeping the riot shield firmly affixed to their front for protection. The gunner takes a step forward and fills the gap. He then leans into the riot shield bearer's back with his knee and fires. This technique ensures that the long end of the weapon is extended beyond the riot shield.

WARNING
The long end of the weapon must be projected beyond the riot shield before firing. This is a critical safety step, which will prevent NL projectiles from hitting the riot shield and injuring soldiers.

4-69. Once soldiers firing munitions have fired, they raise their weapon back to a high port arms. The soldiers clear their weapons and tap the riot shield bearer a second time, signaling them to rise to a standing position.

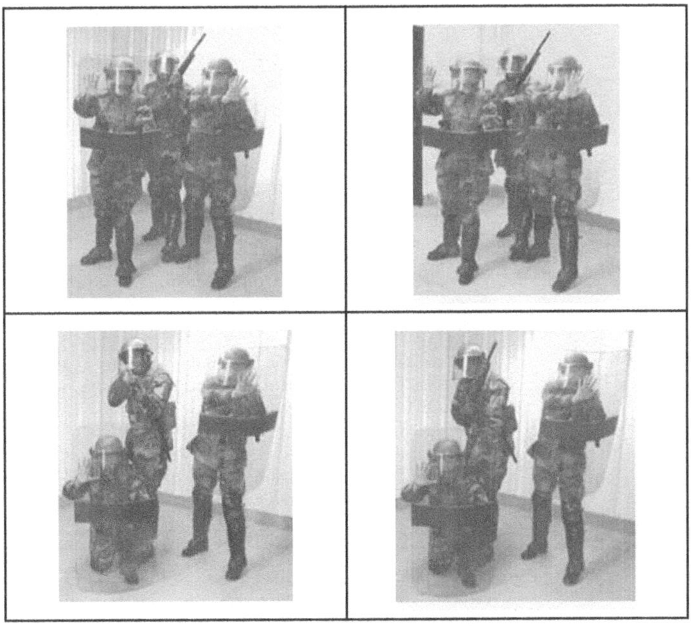

Figure 4-21. Tap-Down Technique

Chapter 5

Nonlethal Capabilities Set and Employment Considerations

During the past several decades, US forces have regularly performed peacekeeping and peace enforcement missions at a greater tempo than in the past. In the peacekeeping and peace enforcement environment, the use of conventional firearms or the threat of their use may not be the solution to a situation where US forces must separate two belligerent, hostile ethnic groups or prevent a similar group from entering an area that is off-limits to them. However, a show of force using NLW and NL munitions will cause crowds to disperse, separate, or leave the area with minimal casualities. The NLCS is a well-rounded, versatile package of both commercial off-the-shelf (COTS) and government off-the-shelf (GOTS) NL equipment and munitions. NLCS contents are divided into four distinct categories: personnel protectors, personnel effectors, mission enhancers, and training devices.

PERSONNEL PROTECTORS

5-1. Personnel protectors provide the individual soldier with added protection to the sensitive and vital areas of his body. They provide excellent protection for the individual soldier from trauma often inflicted by thrown objects (such as rocks, bricks, sticks, and bottles).

NONBALLISTIC RIOT FACE SHIELD

5-2. A nonballistic riot face shield (see *Figure 5-1,* page 5-2) has two adjustable positions and is constructed of hard durable plastic that securely attaches to the Kevlar® helmet. It may be worn with an M17 or M40 protective mask and is designed to protect the soldier's head, face, and neck from thrown objects. Nonballistic riot face shields will not protect from attacks with ballistic weapons, such as firearms. All soldiers operating in a civil disturbance environment should be outfitted with riot face shields.

NONBALLISTIC FULL-LENGTH SHIELD

5-3. A nonballistic full-length riot shield (see *Figure 5-2,* page 5-2) is a body shield constructed of hard durable plastic. Left- and right-handed personnel can easily use the nonballistic full-length riot shield. It is designed to protect the soldier's face, torso, and upper legs from thrown objects. Nonballistic full-length riot shields will not provide protection from ballistic weapons. Only the front line of skirmishers and those providing lateral protection to both flanks of the formation will be outfitted with nonballistic full-length riot shields.

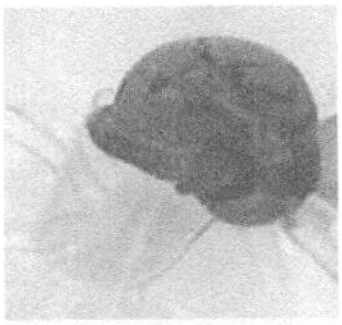

Figure 5-1. Nonballistic Riot Face Shield

Figure 5-2. Nonballistic Full-Length Riot Shield

NONBALLISTIC SHIN GUARD

5-4. The nonballistic shin guards (see *Figure 5-3*) are constructed of hard durable plastic and are easily fastened to a soldier's legs. They are designed to protect the soldiers, feet, shins, and knees from thrown objects. They will not provide protection from ballistic weapons, such as firearms. All soldiers operating in a civil disturbance environment should be outfitted with shin guards.

BALLISTIC RIOT FACE SHIELD

5-5. The ballistic riot face shield (see *Figure 5-4*) has two adjustable positions constructed of acrylic and bullet-resistant materials that provide the wearer with Level IIIA protection. The ballistic riot face shield easily attaches to the Kevlar helmet and is capable of being worn with either an M17 or M40 protective mask. It will protect the soldier from hand-thrown objects. Because of its excessive weight (3.4 pounds), the ballistic riot face shield should not be worn for standard riot control operations. Ballistic riot face shields are best suited for use within specialized teams associated with riot control operations; for example, teams responsible for the search and apprehension of identified criminals or building- and alley-clearing operations.

Figure 5-3. Nonballistic Shin Guards

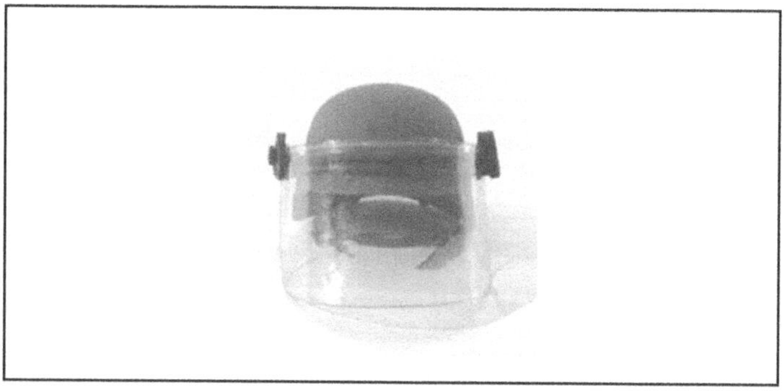

Figure 5-4. Ballistic Riot Face Shield

BALLISTIC FULL-LENGTH RIOT BODY SHIELD WITH LIGHT KIT

5-6. The ballistic full-length riot body shield (see *Figure 5-5,* page 5-4) is constructed from Spectra Shield® bullet-resistant material that provides a soldier with Level IIIA protection. It will protect the soldier from thrown objects. Because of its excessive size (18 pounds and 20 by 36 inches), the ballistic full-length riot body shield should not be used as part of a standard riot control operation. Ballistic full length riot body shields are best suited for use within specialized teams associated with riot control operations; for example, teams responsible for the search and apprehension of identified criminals or building- and alley-clearing operations.

Figure 5-5. Ballistic Full-Length Riot Body Shield

BALLISTIC SHIN GUARDS

5-7. Ballistic shin guards (see *Figure 5-6*) are constructed of Kevlar KM2 materials and provide a soldier with Level IIIA protection to the feet, shins, and upper legs. They protect the soldier from hand-thrown objects. There are three different sizes, and because of their excessive weight (7 to 10 pounds), ballistic shin guards should not be deployed as part of a standard riot control operation. Ballistic shin guards are best suited for use within specialized teams associated with riot control operations; for example, teams responsible for the search and apprehension of identified criminals or building- and alley-clearing operations.

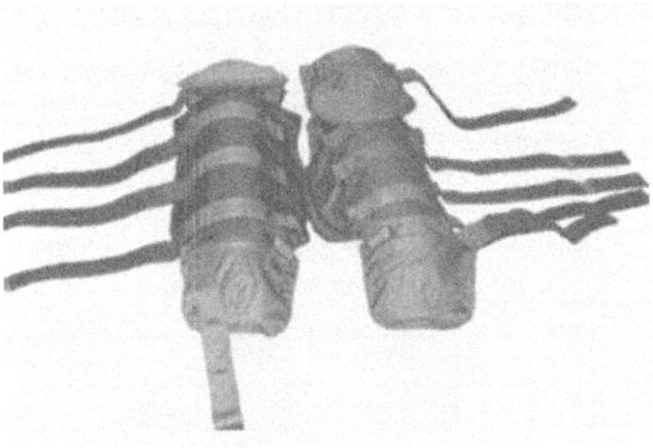

Figure 5-6. Ballistic Shin Guards

PERSONNEL EFFECTORS

5-8. Personnel effectors are those items that provide soldiers with an NL response without the necessity of always coming into direct physical contact with aggressors. Personnel effectors are 12-gauge and 40-millimeter ammunition that inflict blunt trauma. These munitions are compatible with the shotgun and the M203 grenade launcher already in the Army inventory. There are also diversionary and rubber ball grenades, which distract and inflict blunt trauma.

WOOD RIOT BATON WITH BELT RING (36 INCHES)

5-9. The 36-inch wood riot baton (see *Figure 5-7*) is made from solid oak. When held properly, the riot baton thong provides greater retention capabilities. Intended primarily for self-defense, the 36-inch wood riot baton can be used effectively as an offensive tool if it becomes necessary to keep rioters out of arm's reach of soldiers conducting crowd control operations.

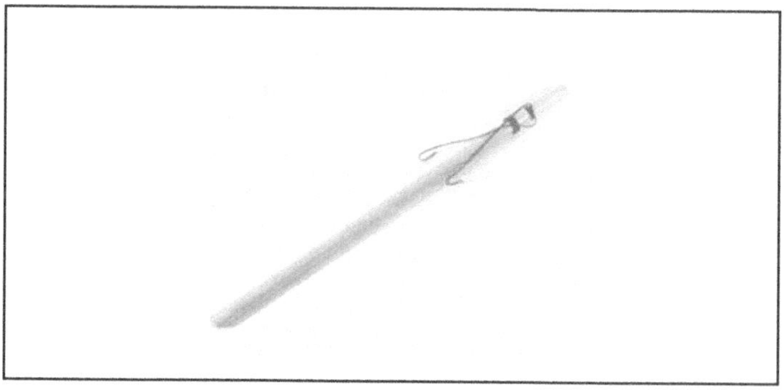

Figure 5-7. Wood Riot Baton

EXPANDABLE RIOT BATON WITH CARRIER (24 TO 36 INCHES)

5-10. The expandable riot baton (see *Figure 5-8*) provides the soldier with an NL means of crowd control and self-defense. The riot baton comes with a mounting device that attaches the riot baton to a belt. The riot baton is 24 to 36 inches in length and is primarily intended for self-defense. The expandable riot baton can be used effectively as an offensive tool if it becomes necessary to keep rioters out of arm's reach of soldiers conducting crowd control operations.

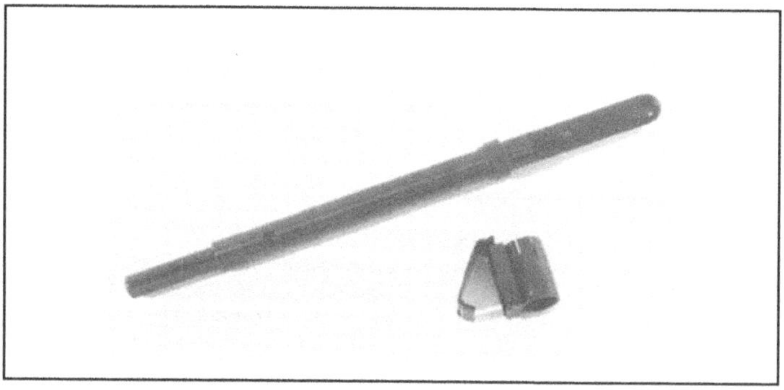

Figure 5-8. Expandable Riot Baton and Carrier

500 (3 INCH) 12-GAUGE SHOTGUN

5-11. The 12-gauge shotgun (see *Figure 5-9*) is a pump action shotgun currently in NLCS inventory. The pump action shotgun is chambered to take up to 3-inch shells. The 3-inch chamber allows for the use of M1012 and M1013 NL munitions. This shotgun also provides a visually distinct alternative to standard military weapons that may be desired based on mission considerations (force signature and acceptability).

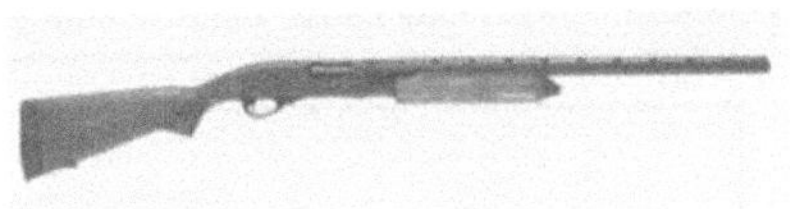

Figure 5-9. 12-Gauge Shotgun

M7 66-MILLIMETER LAUNCHER WITH M315 INSTALLATION KIT

5-12. The M7 (see *Figure 5-10*) is a 66-millimeter vehicle-mounted, NL, grenade-launching device that is mounted on a HMMWV. It is an indirect fire support system that can deliver the M98 distraction grenade that creates a flash-bang effect, L96A1 antiriot grenade, or M99 blunt trauma grenade that creates a sting-ball effect. The M315 installation kit is used to install an M7 discharger on the turret ring of appropriate HMMWV variants. An adjustable bracket allows the launch angle to be depressed for engaging targets at ranges of 50, 75, and 100 meters. The system enforces standoff distances and deters potential threats.

Figure 5-10. M7 Launcher

MODIFIED CROWD CONTROL MUNITION-GROUND EMPLACEMENT

5-13. Similar in nature to the claymore mine, the modified crowd control munition-ground emplacement (MCCM-GE) (see *Figure 5-11*) is an NL munition that can deliver six hundred .32-caliber rubber pellets and a flash-bang effect. The MCCM-GE will stop, confuse, disorient, and/or temporarily incapacitate area targets and/or personnel at ranges out to 30 meters.

WARNING
Avoid employing the MCCM-GE within 10 meters of personnel as it may cause serious bodily harm or death.

Figure 5-11. MCCM-GE

12-GAUGE, FIN-STABILIZED POINT, NONLETHAL CARTRIDGE (M1012)

5-14. The M1012 (see *Figure 5-12*) is a single projectile round made of hard rubber that is shaped like a bomblet and designed to be fired a at a single target. With a muzzle velocity of 500 feet per second, the M1012 has the effective range of no closer than 5 meters and no further than 30 meters. Engagement inside 5 meters could cause serious injury or death. Beyond 30 meters the kinetic energy dissipates to the point where the round becomes ineffective.

Figure 5-12. M1012

12-GAUGE, CROWD DISPERSAL, NONLETHAL CARTRIDGE (M1013)

5-15. The M1013 (see *Figure 5-13*) is a multiple projectile round with .23 caliber hard rubber pellets that is designed to be fired at and employed with the purpose of affecting multiple targets. With a muzzle velocity of 900 feet per second, the M1013 has an effective range of no closer than 5 meters and no further than 30 meters. Engagement inside 5 meters could cause serious injury or death. Beyond 30 meters, the kinetic energy dissipates to the point where the rubber pellets become ineffective.

Figure 5-13. M1013

40-MILLIMETER SPONGE GRENADE (M1006)

5-16. The M1006 (see *Figure 5-14*) is a single projectile round made of pliable foam rubber with a hard plastic back. The M1006 is designed to be aimed and fired at a single target. With a muzzle velocity of 265 feet per second, the M1006 has an effective range of no closer than 10 meters and no further than 50 meters. Engagements inside 10 meters could cause serious injury or death. Beyond 50 meters, the kinetic energy dissipates to the point where the round becomes ineffective.

Figure 5-14. M1006

40-MILLIMETER CROWD DISPERSAL CARTRIDGE

5-17. The crowd dispersal cartridge (CDC) (see *Figure 5-15*) is a multiple projectile round with .48 caliber hard rubber balls that is designed to be fired and employed with the purpose of affecting multiple targets. With a muzzle velocity of 450 feet per second, the CDC has an effective range of no closer than 10 meters and no further than 30 meters. Engagement inside 10 meters could cause serious injury or death and beyond 30 meters the kinetic energy dissipates until the rubber balls become ineffective.

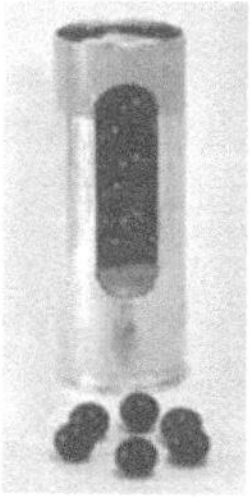

Figure 5-15. CDC

RUBBER BALL GRENADE (GG04)

5-18. The GG04 (see *Figure 5-16*, page 5-10) is a multiple projectile, flash-bang grenade with 100 .25-caliber hard rubber pellets. Each grenade has a fuse delay of 2.8 to 3 seconds with a flash measuring approximately 1 million CP and 180 decibels at 3.5 feet. At detonation, rubber pellets are dispatched at 360° with an effective range of 2 to 3 meters and a maximum engagement range of 15 to 20 meters. The rubber ball grenade is designed to be hand-thrown or muzzle-launched from a 12-gauge shotgun.

RUBBER BALL GRENADE LAUNCHING CUP AND AA30 12-GAUGE LAUNCH ROUND

5-19. The launch cup (see *Figure 5-17*, page 5-10) firmly attaches to the 12-gauge shotgun without any additional tools. It works in concert with the AA30 12-gauge launch round to propel the rubber ball grenade approximately 100 meters. The shooter simply inserts the full body and the safety lever of the grenade into the launching cup, pulls the pin, loads a 12-gauge round, and fires (approximately) at a 30° angle. The desired effect is to have the grenade explode approximately 7 to 10 feet above the target.

RELOADABLE FLASH-BANG GRENADE (M84)

5-20. The M84 (see *Figure 5-18*, page 5-10) is a reloadable flash-bang thrown grenade. Each grenade has a 1- to 2-second fuse delay, with a flash capability of 1.5 to 2.5 million CP and a bang of 168 to 175 decibels. Although it is part of the NLCS, use of the M84 is not recommended for use in crowd control situations. Specialty teams, such as security response teams, and search teams may use this device effectively.

Figure 5-16. GG04

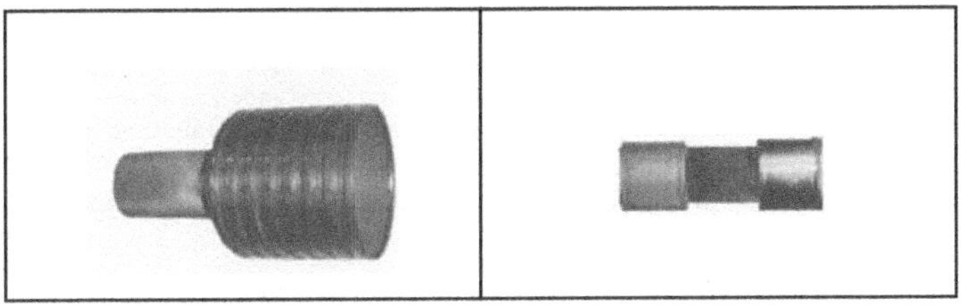

Figure 5-17. Launching Cup and AA30

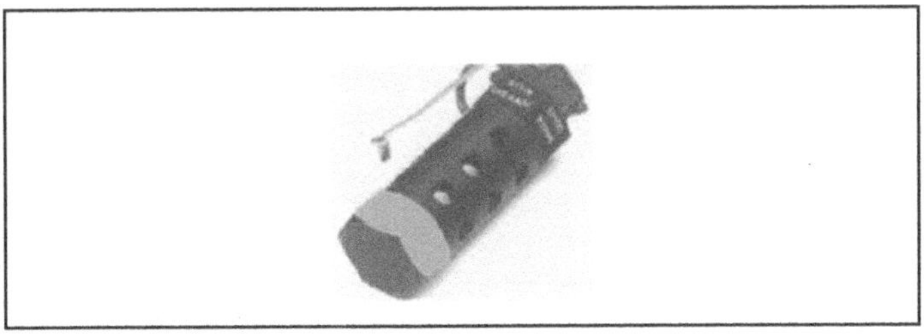

Figure 5-18. M-84

FLEX-CUFS

5-21. Flex-Cufs consist of a tough pliable plastic band with a self-locking mechanism in the center for each end. When the Flex-Cuf (see *Figure 5-19*) is threaded the restraint bands circle around the wrists or ankles, impeding movement and securing the individual. They are lightweight and disposable and require a cutting tool or other instrument to remove.

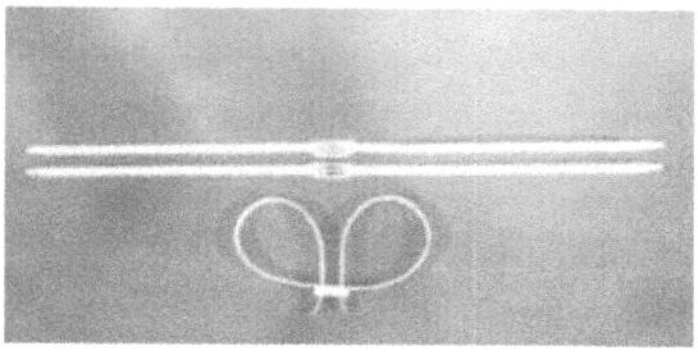

Figure 5-19. Flex-Cufs

INDIVIDUAL RIOT CONTROL AGENT DISPERSER (M36)

5-22. The M36 (see *Figure 5-20*) contains a Dibenz (F, -1 4-oxazepine [CR] solution. It can deliver 25 one-second bursts out to 12 feet. Individual disposable RCA dispersers are intended primarily for self-defense or to keep rioters out of arm's reach of soldiers conducting crowd control formations or soldiers engaged in missions where a noncombatant exists.

Figure 5-20. M36

MIDSIZE RIOT CONTROL AGENT DISPERSER (M37)

5-23. The M37 (see *Figure 5-21,* page 5-12) is the size of a standard fire extinguisher that uses compressed air to force the RCA out to a range of 30 feet. It has the capacity to employ 18 bursts for 3 seconds per charge. It is excellent for providing a wide coverage of RCAs onto a hostile crowd while maintaining excellent standoff capability. The M37 can be refilled and is rechargeable. It can be filled with CR solution (liquid agent) or CS1 (dry agent). For training purposes, CR can be substituted with water and CS1 can be substituted with talcum powder.

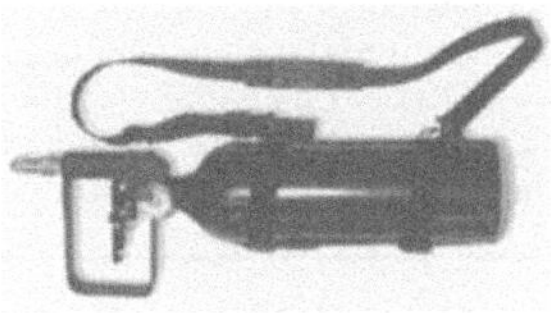

Figure 5-21. M37

SQUAD RIOT CONTROL AGENT DISPERSER (M33A1)

5-24. The M33A1 (see *Figure 5-22*) is designed to provide crowd control and protection at the squad level. It is capable of projecting a ballistic stream of RCAs beyond 25 feet in up to 25 half-second bursts. It consists of a frame and harness assembly, compressed-gas cylinder (agent container assembly), air pressure assembly, gun and hose assembly, multijet spray unit, and check valve assembly. The M33A1 can be refilled and is rechargeable. For training purposes, CR can be substituted with water and CS1 can be substituted with talcum powder.

Figure 5-22. M33A1

MISSION ENHANCERS

5-25. Mission enhancers aid the commander in various tactical situations. Mission enhancers include various types of devices used for illumination, vehicle denial, traffic control missions, and force protection enhancement.

PORTABLE BULLHORN

5-26. The portable bullhorn (see *Figure 5-23*) provides squad leaders with the capability to project their voices above the noise and commotion created by crowds and mobs. The megaphone has a built-in siren for alarm purposes that allows the operator to use the device as an ordinary megaphone and siren alarm during emergencies. The bullhorn is portable and easy to carry. The acoustic range of the megaphone for a clear voice output is ¾ to 1 mile. The power output of the megaphone is rated at 15 watts with a maximum rating of 20 watts. The dimensions of the megaphone are 14.5 inches with a 9-inch diameter bell. It weighs 3.5 pounds without batteries. Battery requirements for the megaphone are eight dry-cell size C batteries.

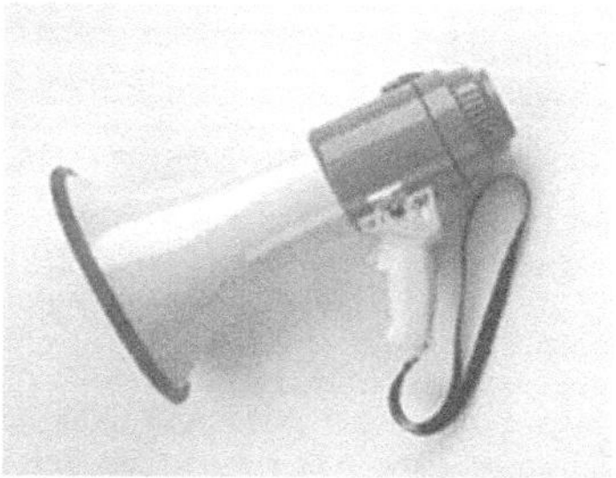

Figure 5-23. Portable Bullhorn

GROUND-MOUNTED BULLHORNS

5-27. Ground-mounted bullhorns (see *Figure 5-24,* page 5-14) are a critical communication enhancement device for conducting crowd control tactics. The bullhorn can facilitate communication with the crowd in conjunction with PSYOP support and assistance with the communication of commands to troops engaged in the crowd control process. The power output of the ground-mounted bullhorn is rated at 22 watts. The dimensions of the bullhorn are 13.75 by 9 by 14.75 inches, and it weighs 15 pounds.

INDIVIDUAL VOICE AMPLIFICATION SYSTEM (M7)

5-28. The M7 (see *Figure 5-25,* page 5-14) is a critical communication enhancement device for conducting crowd control tactics using RCAs while wearing the M40 protective mask. The M7 is fitted to the M40 protective mask to facilitate oral communication and increase the user's ability to communicate using radios and other devices.

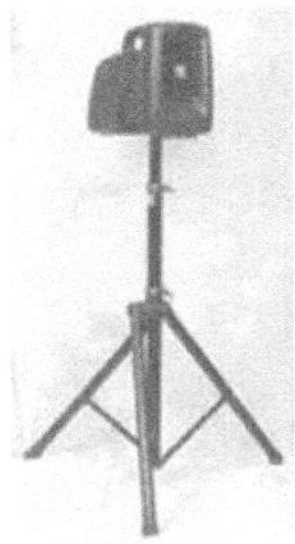

Figure 5-24. Ground-Mounted Bullhorns

Figure 5-25. Individual Voice Amplification System

INDIVIDUAL HIGH-INTENSITY SEARCHLIGHT WITH HOLSTER

5-29. The individual high-intensity searchlight (see *Figure 5-26*) is used primarily for illumination in crowd control operations at night. An individual can use it for general illumination of the operational area to pinpoint agitators and threats, reduce the ability of the rioters to see troop formations and actions, enhance tactical deception and actions, and enhance tactical deception techniques for units conducting crowd control operations.

HIGH-INTENSITY LIGHT

5-30. A high-intensity light (see *Figure 5-27*) is intended for use in low light or night conditions. A high-intensity light can project a beam that will enable the user to identify an individual person up to 1,900 yards away. It has an adjustable beam spread of 1° to 15° and will run continuously at maximum power for up to 45 minutes. The internal power supply requires 4 hours for a full charge.

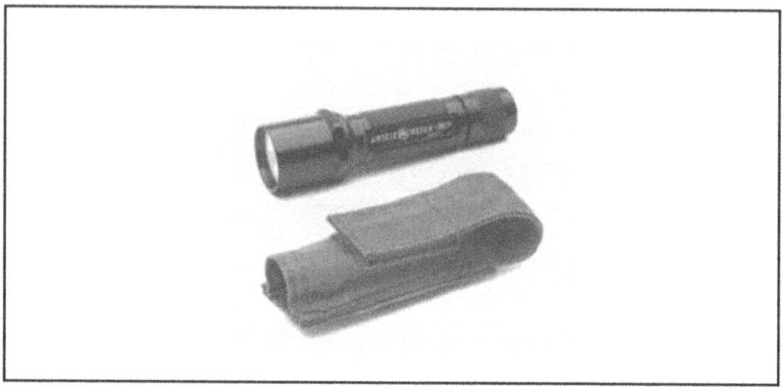

Figure 5-26. Individual High-Intensity Searchlight with Holster

Figure 5-27. High-Intensity Light

HIGH-INTENSITY LIGHT SYSTEM WITH 12-GAUGE SHOTGUN ATTACHMENT KIT

5-31. This 12-guage high-intensity light system (see *Figure 5-28,* page 5-16) is used to illuminate possible targets in crowd control situations where there is little or no light. It serves to blur or blind the aggressor's ability to see. It mounts easily to the end of the 12-gauge shotgun barrel.

40-MILLIMETER CARRYING POUCH

5-32. The 40-millimeter carrying pouch is made from durable canvas material, (see *Figure 5-29,* page 5-16) and can be easily affixed to load-bearing equipment (LBE) and/or a load-bearing vest (LBV) using Velcro straps. It is capable of carrying up to six 40-millimeter rounds.

Figure 5-28. 12-Gauge High-Intensity Light System With 12-Gauge Shotgun Attachment Kit

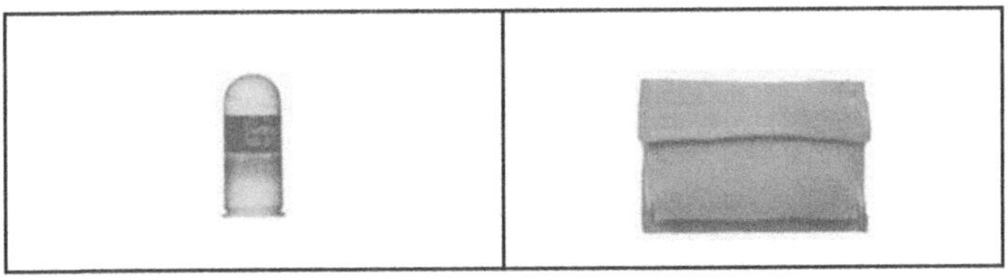

Figure 5-29. 40-Millimeter Carrying Pouch

12-GAUGE UTILITY POUCH

5-33. The 12-gauge utility pouch is made from durable canvas material (see *Figure 5-30*). It can be easily affixed to an LBE and/or LBV using metal clasps. The 12-gauge utility pouch is capable of carrying up to twenty-five 12-gauge shotgun rounds.

Figure 5-30. 12-Gauge Utility Pouch

12-GAUGE BUTTSTOCK CUFF

5-34. The 12-gauge buttstock cuff (see *Figure 5-31)* firmly attaches to the shotgun by sliding over the stock. Elastic straps aid in preventing the buttstock cuff from sliding or rolling. It holds up to seven 12-gauge cartridges, which give the soldier quick access to additional rounds.

Figure 5-31. 12-Gauge Buttstock Cuff

DIVERSIONARY AND/OR RUBBER BALL GRENADE POUCHES

5-35. The diversionary and/or rubber ball grenade pouch (see *Figure 5-32*) is an ammunition-carrying pouch made of durable canvas material. It is partitioned in a manner to provide the soldier with a means of holding up to six rubber ball grenades or six flash-bang grenades in separate compartments.

Figure 5-32. Diversionary and/or Rubber Ball Grenade Pouch

CALTROPS

5-36. The caltrop (see *Figure 5-33,* page 5-18) is a field-expedient area denial system. Securing approximately 50 caltrops to 550-type military cord allows a soldier to quickly toss and recover the system. Caltrops are used to deny vehicle and pedestrian movement in designated areas. The four prongs are approximately 2 inches tall and will puncture tires or boots. A soldier can easily employ the caltrop by scattering it in a designated area hidden by camouflage.

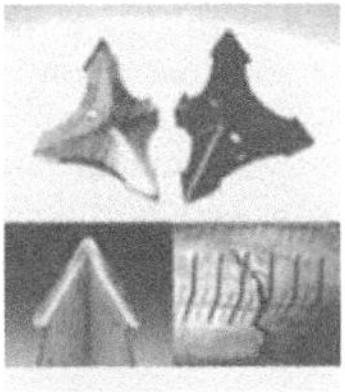

Figure 5-33. Caltrop

PORTABLE VEHICLE ARREST BARRIER

5-37. A portable vehicle arrest barrier (PVAB) (see *Figure 5-34*) is a lightweight, portable, manually-emplaced and recoverable-barrier entrapment device that is used (on command) to create a barrier that will stop vehicles being driven by known or suspected hostile forces with minimal damage to the vehicle and its occupants. Two individuals can easily emplace it in approximately 4 to 8 hours. It has the capability of stopping a light truck (up to 7,500 pounds) at speeds of up to 45 miles per hour with a stopping distance of less than 200 feet. It is not to be used as a stand-alone system. The PVAB may be used at dismount points, checkpoints, and roadblocks to prevent unauthorized access by wheeled vehicles into or out of areas.

Figure 5-34. PVAB

TRAINING DEVICES

5-38. Training devices that commanders can use to effectively prepare their soldiers are included in the NLCS. Training devices allow soldiers to learn and build confidence in the development of personal defensive skills. These are critical pieces of equipment because they allow soldiers to train at almost full speed, which provides reinforcement of realistic motor skills. Training in this manner has proven to be more effective than just striking out at the air.

IMPACT TRAINING SUIT

5-39. Impact training suits (see *Figure 5-35*) are made of closed-cell shock-absorbent foam that allows soldiers to hone their riot baton skills through realistic hand-to-hand training. Impact training suits absorb strikes inflicted by the current expandable and wooden riot batons. They provide protection for the head, face, hands, and legs without significant degradation of the wearer's mobility.

Figure 5-35. Impact Training Suits

TRAINING STRIKE BAGS

5-40. Training strike bags (see *Figure 5-36*) give the soldier a durable, yet absorbing surface to strike or kick for training. It also helps develops proficiency in open-hand control and riot baton techniques.

Figure 5-36. Training Strike Bag

PRACTICE GRENADE BODY (GG05)

5-41. The GG05 is blue in color and is easily identified as a practice grenade (see *Figure 5-37*). It is designed to replicate the live GG04 rubber ball grenade. It is made of durable rubber and can be hand-thrown or muzzle-launched from the 12-gauge shotgun. It can be reloaded.

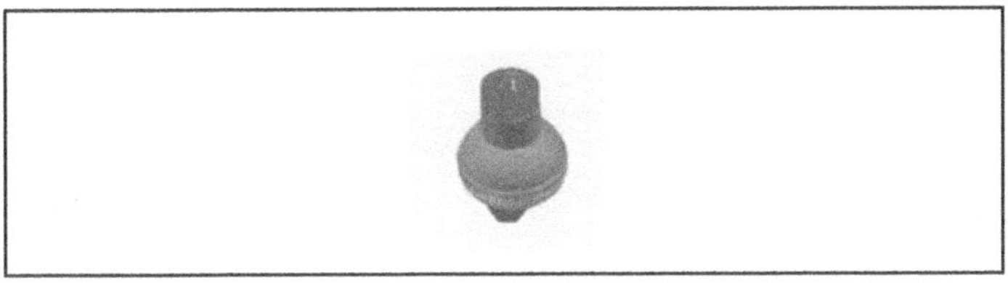

Figure 5-37. GG05

PRACTICE GRENADE FUSE

5-42. The practice grenade fuse (see *Figure 5-38*) is identified by its blue spoon and brown stripe. It easily screws into the body of the GG05. Once used, unscrew the practice grenade fuse from the body and reload it with a new fuse.

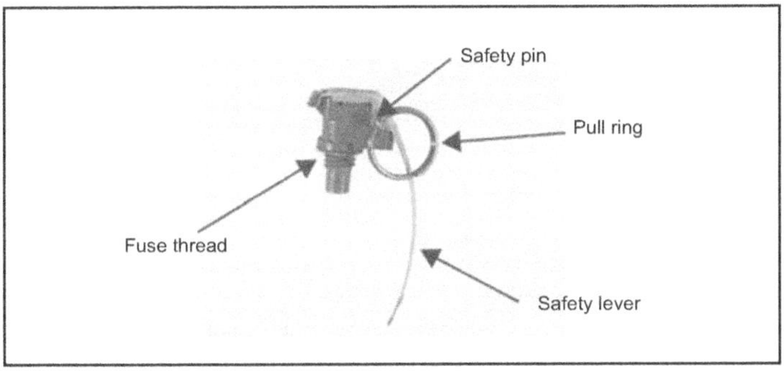

Figure 5-38. Practice Grenade Fuse

CORE CAPABILITIES

5-43. Core capabilities are those fundamental competencies that enable the commander to achieve the desired operational outcome. In the case of NLW, this includes providing a flexible means of response in order to protect friendly forces and/or influence the actions of potential adversaries and noncombatants. With the availability of the NLCS (see *Figure 5-39*) these goals can be achieved without resorting to lethal force and in a manner that will minimize collateral damage. The core capabilities associated with NL effects fall into the following two major categories:

NOTE: There are nine platoon equivalents within a light infantry battalion.

PLT Organization	Personnel and Weapons	Nonlethal Weapons & Equipment
PLT	7 personnel (5 M16/M4 and 2 M60/ M240)	3 riot face shields 3 sets of shin guards 3 sets of full-body restraints 3 individual voice amplification systems 3 individual RCA dispensers with carriers 1 portable bullhorn 1 ground-mounted bullhorn 1 high-intensity searchlight 1 set of caltrops 3 ballistic riot face shields 3 sets of ballistic shin guards 3 ballistic riot body shields 1 riot baton training suit 1 riot baton strike pad
SQD SQD SQD	3 personnel (3 M16/M4)	3 riot face shields 3 sets of shin guards 3 riot body shields 3 expandable riot batons with carriers 9 sets of cinch-strap restraints 3 individual voice amplification systems 3 individual RCA dispensers with carriers 3 individual high-intensity search lights with carriers (see *Figure 5-26*)
TM TM TM TM TM TM	24 personnel (12 M16/M4 and 6 M16/ M4 with M203 6 SAW)	24 riot face shields 24 sets of shin guards 100 sets of disposable flex-cuffs 24 individual voice amplification systems 24 individual RCA dispensers with carriers 6 expandable riot batons with carriers 11 grenade pouches 12 forty-millimeter grenade pouches 6 twelve-gauge ammunition pouches 6 twelve-gauge bitstock ammunition carriers

Figure 5-39. NLCS Distribution

- **Counterpersonnel.** NL counterpersonnel capabilities enable the application of military force with a reduced risk of fatalities or serious casualties among noncombatants or, in some instances, among enemy forces. There are several specific NL counterpersonnel capabilities to be explored. These include the means to influence the behavior and activities of a potentially hostile crowd, and the capability to bring a mob engaged in a riot under control. While there are many similarities in these two scenarios, each involves unique challenges, which may require radically different solutions. For more information, see *FM 3-22.40.*
- **Countermaterial.** NL countermateriel capabilities enhance operations by rendering equipment and facilities unusable without complete destruction. An NL countermateriel capabilities enable the

application of military force to defuse potentially volatile situations under circumstances in which more destructive conventional military means might prove counterproductive. For example, pre-emptive strikes against troublesome and aggressive nations may be politically unacceptable when only conventional weapons with the high risk of casualties are involved. With NL countermateriel capabilities, the enemy's ability to threaten its neighbors could be curtailed with less political risk by attacking only their weapons of war and their supporting infrastructure. For more information, see *FM 3-22.40*.

ADDITIONAL CAPABILITIES

5-44. NLW are only NL when fired within their parameters.

DANGER
If NLW are fired outside their parameters, they may cause serious injury or death.

5-45. If a situation turns from NL to lethal, an NL gunner only needs to change his point of aim and distance to inflict serious bodily harm and/or death to the threat. The closer the soldier is to the target, the greater the kinetic energy and the more severe the effects are to the threat.

EMPLOYMENT CONSIDERATIONS

5-46. NL munitions and equipment are not completely NL; they are NL by intent. The term NL does not guarantee zero mortality or nonpermanent damage. NLW can add flexibility to combat operations and enhance force protection by providing an environment in which friendly troops can engage threatening targets with a reduced risk of noncombatant casualties and collateral damage.

5-47. If the tactical situation dictates a NL response, a soldier who levels a weapon loaded with NL munitions must be trained in distance to the target and appropriate points of aim. Based on the soldier's understanding and training on the characteristics and capabilities of the round, they can easily discern and engage targets to the desired effect. If the tactical situation dictates a lethal response, soldiers can easily adjust the point of aim to a more vulnerable target location on the body and move in closer to the target. NL munitions can be used as a lethal response if the situation dictates and the intent of the soldier is to inflict death.

NONLETHAL WEAPONS CAPABILITIES IN FORMATIONS

5-48. The mix of NLW within the formation is based on METT-TC. For example, the first rank of the formation will have riot shields, riot batons, and their personal weapon (slung across their back [left to right] with the butt up and muzzle down or holstered). The second row will have a mix of shotguns and M203s. Shotguns and M203s may be used as area denial or point target weapons depending on the specific munitions used. The shotgun provides cover for the slower loading M203. The commander may move these weapons within the formation to meet the needs of the mission and to create distance between the formation and the crowd (see *Appendix D*).

Chapter 6

Civil Disturbance Formations

Control force formations, when properly employed and executed against a crowd, are some of the most practical methods of crowd control. The riot baton has been the weapon of choice for the control force. With the development of NLW, the commander now has more options at his disposal. Before NLW, the only other choices were chemical irritants, MWDs, and lethal force. Now commanders can create a standoff distance with NLW that reduces the risk of close fighting with the mob. During any crowd control operation, the commander must protect the control force with lethal overwatch. This overwatch may be in the form of marksmen and observer teams (sniper with observer) and a response force with lethal capabilities.

OVERVIEW

6-1. With advances in NLW, it is possible to create an effective standoff distance of 15 to 100 meters between the crowd and the control force formation. It is now possible to advance and disperse the crowd without coming into direct contact with it. Some NLW may also be used to assist in the identification and apprehension of instigators.

6-2. Control force formations are used to disperse, contain, and/or block a crowd. These formations are more effective in urban areas than in open areas, such as parks or fields. When control force operations are employed in urban areas, it is easier to split a crowd into smaller segments, isolate instigators, or funnel the crowd into the desired location by using buildings and other man-made structures. Commanders must be mindful of the fact that well-organized crowds may attempt to move to the flanks of a formation and/or get behind the formation to gain a tactical advantage.

6-3. Commanders must be aware of the limitations of control force formations. They are not the answer to all civil disturbance situations. Do not expose the formation to sniper fire and/or unnecessary violence simply for a show of force. Control force formations should be used when it is decided that a crowd poses a threat and must be dispersed or moved to a specified area. Before dividing a large crowd, the commander must consider that this might not solve the problem. It may worsen the problem by creating smaller elements that may engage the control force in small-mob tactics, such as sniping, looting, burning, and attempting to envelop the control force. Commanders must anticipate that breakaway crowds may engage in small-mob tactics. These tactics can be defeated by area control measures, such as building searches, saturation patrolling, and other tactics. After dispersing the crowd, the control force must not assume that there will be an immediate return to peaceful activities. The use of formations is only part of the total dispersal effort.

6-4. Commanders must assess and secure the areas of operations (AOs). This may be accomplished with the use of helicopter overflights and by visually securing rooftops, high elevations, and advantage points. Designated marksmen teams may be used to secure the rooftops and provide lethal overwatch for the control formation. The control force commander must be aware that designated marksmen teams are positioned on rooftops.

6-5. Crowd control formations and their support teams have more capabilities than just crowd dispersal. They also have the capability to apprehend and detain certain members of the crowd that the commander feels may instigate the crowd to further violence. Commanders must analyze the threat, determine the mission of the control force, and decide which control force formations will accomplish the mission according to METT-TC.

6-6. During CONUS operations, there are usually no enemy forces. However, commanders must analyze and prepare for any potential opposition. The formations discussed in this chapter are guidelines and may be adapted to fit any mission or situation. Whatever the modification, soldiers must stay in position and on line. Only through training and rehearsals will soldiers become proficient in basic formation movements and be able to adjust to changing situations. Soldiers must also be prepared to employ NLW. *Figure 6-1* is used as a key to identify soldiers within crowd control formations.

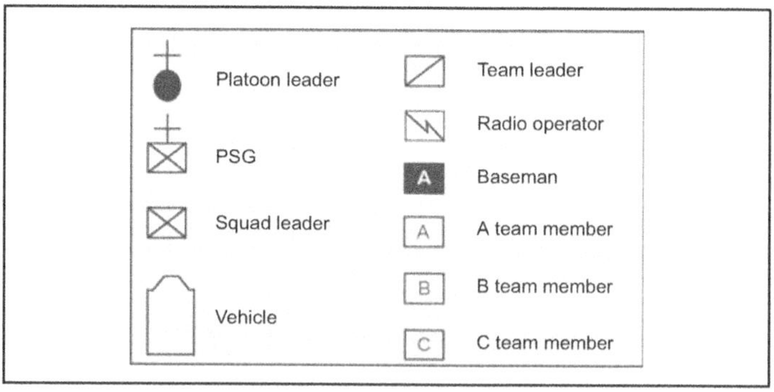

Figure 6-1. Symbols for a Formation Element

ACTIONS BEFORE MOVEMENT

6-7. Before moving the control force to the civil disturbance area and into a potentially hostile urban environment, the commander and his staff must review the preincident plans and expand them (as needed) based on current information. Because of the employment of forces against ambiguous threats,

IPB becomes critical. Information is the key to developing an appropriate response to civil disturbances.

6-8. Civil disturbance training must be part of the annual unit training plan. If a soldier is deployed to an area where civil disturbances are a recurring event, sustainment training must occur more frequently. Before moving the control force to the rally point, an intense rehearsal must be conducted by the leadership and the control force. A top-down review of the rules of engagement (ROE/rules for the use of force (RUF) must be accomplished. The ROE/RUF must be part of the unit training plan and each soldier must be trained to standard (see *Chapter 8*). The use of NLW (type specific) must be granted to the on-site commander from the approval authority.

6-9. Information is the key to developing plans for appropriate responses to a civil disturbance, and IPB is crucial to the success of the mission. During the planning phase (see *Chapter 2*), the information provided by the IPB will aid the staff in selecting various routes and rally points. The selection of ingress and egress routes must include a variety of ways in and out of the area. Direct and indirect routes and those with cover and concealment are necessary. The rally point must be cleared and secured by an advanced party capable of controlling the area.

RALLY POINT ACTIONS

6-10. Once at the rally point, the commander makes contact with the local police, civil authorities, or military officer. He then assesses the situation, decides the next COA, and issues orders, as appropriate. The platoon sergeant (PSG) forms the platoon into a column formation and places team leaders directly in front of their teams (see *Figure 6-2,* page 6-4). Team and squad leaders make last minute checks of their soldiers and await orders.

6-11. The commander places his lethal overwatch (designated marksmen [DM] teams) in areas that overlook the control force and the mob. They may be to the flanks and slightly to the front of the formation. It is their responsibility to protect the formation from lethal fire by constantly scanning the crowd.

6-12. When the commander receives word that DM teams are in place, he forms the control force into a crowd control formation and moves them quickly from the rally point to where the crowd is assembled. Depending on the situation, movement should be in a column formation or one of the three basic control force formations. Reserve forces are left at the rally point until ordered forward as additional lethal overwatch or to reinforce the formation.

CONTROL FORCE FORMATIONS

6-13. Control force formations, when properly employed and executed against a crowd of limited size, are one of the most practical methods of crowd control. Experience has indicated that the most frequently used control force

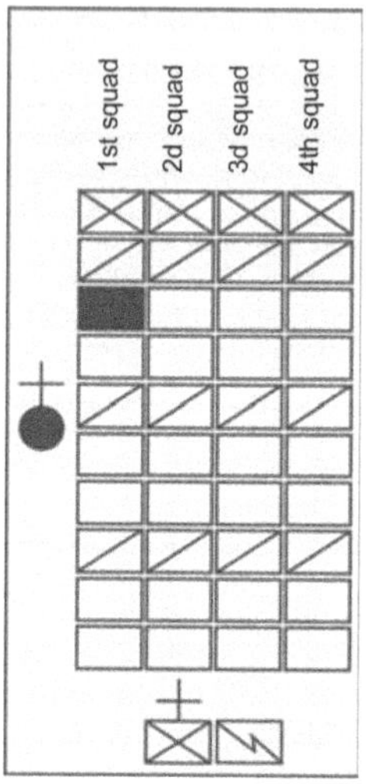

Figure 6-2. Column Formation

formations are the line, echelon, and wedge. The two lesser-used formations are the diamond and circle. Descriptions of all the formations are as follows:

- **Line formation.** The line is the basic formation and it is used more often because of its offensive and defensive applications. As an offensive formation, the line is used to push or drive crowds straight back, across an open area, or up a city street. As a defensive formation, the line is used to hold the crowd or to deny access to areas.
- **Echelon formation.** The echelon is an offensive formation used to turn or divert groups in open or built-up areas and to move crowds away from buildings, fences, and walls.
- **Wedge formation.** The wedge is an offensive formation that is used to penetrate and split crowds into smaller groups.
- **Diamond formation.** The diamond, when used as an offensive formation, is used to enter a crowd and is the formation of choice for extraction teams. As a defensive formation, the diamond is used when all-around security is required in open areas. The decision to use this formation is based on the conformation of the crowd.
- **Circular formation.** The circular formation is used for the same purpose as the diamond formation. The decision to use this formation is based on the conformation of the crowd.

6-14. There are many suitable variations of control force formations that may be employed, but appropriate commands and signals must be created to execute the formations. Because of their somewhat complicated nature and the coordination required for these formations, new variations must be trained, practiced, and rehearsed before they are used in a civil disturbance.

VEHICLES AND FORMATIONS

6-15. Vehicles may be employed with troops in control force formations, especially when a large riotous crowd is on hand. Although vehicles add strength to formations, certain precautions must be taken. Covering the windshield with sturdy, close mesh fencing and the standard safety glass will offer some level of protection to the occupants. Shields and/or mobile barriers may be built by mounting a wooden or metal frame strung with barbed wire across the front of a vehicle. Members of the formation should walk as near to the front corners of each vehicle as possible to keep rioters from attacking the sides and rear of the vehicles.

6-16. Up-armored HMMWVs, ASVs, or armored vehicles may be employed with the control force formation when determined by METT-TC. They present a strong psychological effect and offer protection for the occupants. Wheeled armored vehicles are best because they do not deface the pavement. Whenever vehicles are used in a crowd control formation, the commander must ensure that he is still able to see and control the formation. This may require him to occupy a position in a similar vehicle behind the formation or command subordinate leaders to move away from the front line and into line-of-sight (LOS) control. This will require additional hand-and-arm signals to allow the commander and subordinate leaders to communicate. Commanders can effectively use vehicles as primary communication tools by using horns and lights as signals to organize movement or actions. However vehicles are used, it takes a good measure of training and practice to be effective.

ELEMENTS OF A CONTROL FORCE FORMATION

6-17. Four elements make up the basic crowd control formation. They are as follows:

- **Base element.** The base element is made up of two ranks. The first rank is shield holders, while the second rank contains the NLW. This is the front line of the formation.
- **Support element.** The support element forms up in a column formation behind the base element and has a variety of uses. It may be used to replace base element members as needed, provide lateral or direct support, or perform extraction team operations. The support element helps the base element by performing the following three essential tasks:
    - **General support.** The general-support element is formed from an uncommitted squad in the platoon (generally the 2d squad). When a company is tasked as the control force element, one of the platoons becomes the general-support element. The element is in a column formation centered on and behind the main formation. From this formation, the general-support element can move to lateral or direct support, as needed.
    - **Lateral support.** The lateral-support element is used to protect the flanks of the formation. This is done by moving a set number of teams forward from the general-support element or by using the

end teams in the formation; however, this will make your formation smaller. Once in position, these teams become part of the formation, with their riot shields facing the flanks of the formation.

  - **Direct support.** When direct support is ordered, the general-support element moves forward as they form their own formation. While moving forward, the squad and team leaders from the control force formation step back and allow the riot shield holders from the support element to step behind riot shield holders of the original formation. This allows the support riot shield holders to form up between and behind the riot shield holders of the original formation. Now there are two lines of riot shield holders between the crowd and NLW firers. This formation is the strongest of the formations and requires more planning and practice to master.

- **Command element.** The command element contains several different members. A general configuration for the command element is the platoon leader, the PSG, a radio operator, a video recorder operator (if required), and an interpreter (if required). This element does not have a fixed location within the formation and moves about as needed.
- **Lethal overwatch element.** The lethal overwatch element is a team formed from reserve security forces.

6-18. The reserve support element is not part of the control force formation until it is brought forward from the rally point to join the formation. It remains until released by the formation commander. Lethal overwatch teams are formed from the reserve force. Once lethal overwatch teams are deployed and in position, they are under the control of the formation commander. Communication between the lethal overwatch team and the formation commander is a priority because he approves target selection and engagement.

WEAPON POSITIONS

6-19. At the commander's discretion, members of the control force formation may carry their weapons in any of the positions prescribed in *FM 3-21.5*. However, the primary method of carrying weapons in the control force formation is at the safe-port position (see *Figure 6-3*). It allows the soldier to control both ends of the weapon while moving in and out of the formation and advancing on the crowd.

6-20. High-port position (see *Figure 6-3*) is a position of complete readiness. It is used whenever troops are in contact with a crowd that is showing resistance or not withdrawing. This is a tiring position and is hard to maintain for extended periods. Commanders must rest their troops at every opportunity by using less tiring positions.

Safe-port arms | High-port arms

Figure 6-3. Weapons Positions

COMMANDS

6-21. Commands are given to the control force formation orally or with hand-and-arm signals. Oral commands for the on-guard position are given in one count. All other commands are given in two counts, a preparatory command followed by a command of execution. However, oral commands cannot be relied on completely, so commanders must plan to use hand-and-arm signals too. When vehicles are part of the formation, radios become an alternate method for relaying commands. To improve communication, commanders may use hand-and-arm signals (see *Figure 6-4,* page 6-8) with oral commands. The basic commands used in control force formations are outlined in *Table 6-1,* page 6-9. These commands can be added to or combined to form more complex commands, therefore forming more complex formations.

MOVEMENTS

6-22. Control force formations move in the same manner as regular formations. This allows the commander to more precisely maneuver the formation by commanding the number of steps it should move and in what direction, such as "Five Steps, Forward March." The normal rate of march for entering and leaving a control force formation is double-timing. The half-step march is used when the formation is in direct contact with the crowd. This slows the formation down and allows for better command and control. The squad and team leaders echo preparatory commands and provide cadence while in control force formations.

INTERVAL AND DISTANCE

6-23. Interval is the lateral space between elements. Distance is the space between elements in a column. The usual interval and distance between soldiers in control force formations is 30 inches. The interval and distance can be adjusted based on METT-TC. If the crowd has to be physically pushed back, a close interval is preferred. This allows the riot shields to be overlapped, creating a stronger wall. During peaceful demonstrations, the interval could increase to double arm width, allowing the formation to cover more area and still control the crowd. The interval and distance could change several times during the course of operations before the crowd is dispersed.

Figure 6-4. Hand-and-Arm Signals

Table 6-1. Basic Commands

Commands for	Command	Given by	Purpose
Formations	Platoon On Line	Platoon leader and/or commander	To form a line formation.
	Platoon Wedge	Platoon leader and/or commander	To form a wedge formation.
	Platoon Echelon	Platoon leader and/or commander	To form an echelon formation.
	Platoon Diamond	Platoon leader and/or commander	To form a diamond formation.
	Platoon Circle	Platoon leader and/or commander	To form a circular formation.
	Move	Platoon leader and/or commander	As a command of execution. **NOTE: The platoon leader and/ or commander also identifies the location for the formation by pointing his arm to the desired location for the formation.**
Support	General Support	Platoon leader and/or commander	To place a specified unit in the rear of the base element.
	Lateral Support	Platoon leader and/or commander	To place support elements on the left or right flank of the base element.
	Direct Support	Platoon leader and/or commander	To move support elements forward to strengthen the base element.
	Open	Extraction team leader	To open a space in the formation.
	Up	Extraction team leader	To inform the extraction team to start moving back to the main formation.
Fire	Weapon System Gunners, Number of Rounds, Type of Round, and Prepare to Fire	Platoon leader and/or commander	To ready NL gunners to fire weapons; for example, M203 gunners, three rounds, area target, prepare to fire.
	Shield Down	NL gunners	To tell riot shield holders to get down on one knee and lower the riot shield to the ground.
	Shield Up	NL gunners	To tell riot shield holders to stand and raise the riot shield.
Weapons	Port Arms	Platoon leader and/or commander	To use when not in contact with the crowd.
	High Port	Platoon leader and/or commander	To tell NL gunners what position to hold their weapons in while the formation is in physical contact with the crowd. This allows for rapid targeting and firing of NL munitions.

EXTRACTION TEAM OPERATIONS

6-24. The extraction team is a squad (see *Figure 6-5*) that is generally formed from the support element, but could also be formed from reserve security forces. Once the team is formed, it is OPCON to the formation commander until the team's mission is complete and it returns to the rear of the formation. This team provides the riot control formation with the means to employ NL and lethal cover forward of the formation. It also provides the formation with the ability to remove and detain rioters downed by munitions.

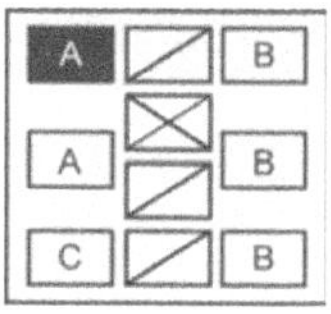

Figure 6-5. Extraction Team Formation

The extraction team—

- Extracts vehicles or personnel that are in immediate danger from the crowd.
- Detains and escorts downed rioters to the rear of the formation.
- Restrains and searches subjects.
- Covers confined or congested areas where a full riot control formation cannot be inserted.

6-25. The extraction team is usually a squad. While the recommended distance for deploying the extraction team from the control force formation is no farther than 10 meters, this may change based on METT-TC. The control formation protects the extraction team and provides immediate support, if necessary.

6-26. The extraction team may deploy from anywhere within the formation. When given orders to deploy, the squad leader of the extraction team gives the formation commands, identifies at least two personnel in the squad to conduct the search and apprehension of the subject, and leads the squad to the base line of the formation. He then extends his arm between the two riot shield holders and commands, "Open," while tapping the riot shield holders on the side of the shoulder. The two riot shield holders take one step backward and one step to the right or left. This clears a path for the extraction team. The extraction team leader then states the number of personnel leaving the formation, such as "Ten security personnel leaving." As the last extraction team member exits the formation he states, "Last Man." The two riot shield holders then return to their original position in the formation. The adjacent base element squad and/or team leaders pass the number of personnel in the team through the formation.

6-27. As the extraction team approaches an instigator or incapacitated demonstrator, the riot shield holders envelop the target and face in the direction of the nearest threat. NL gunners on the left and right cover their respective areas. The search and apprehension team immediately controls and restrains the target. To avoid confusion, one team member concentrates on controlling the subject while another team member applies a restraining system. When the search and apprehension team has the subject under control, the extraction squad leader sounds off with, "Up." Upon hearing the "Up" command, it is repeated by all extraction team members. Team members then grab a shoulder of the team member in front of them to backtrack into the formation. As the extraction team approaches the formation, the nearest base element leader will extend his arm and command, "Open," while tapping the riot shield holders on the side of the shoulder. The riot shield holders move one step backward and to the right or left to open the formation. As he enters the formation, the extraction team leader sounds off with the number of personnel entering the formation, such as "Ten security personnel and one demonstrator." The adjacent base element leader counts the number of personnel as they enter the formation. When the last man enters the formation, he states, "Last Man." The riot shield holders then return to their original places in the formation. The adjacent squad and team leaders in the base element pass information through formation that the extraction team has returned.

LETHAL OVERWATCH TEAMS

6-28. During a NL engagement, the use of a DM team provides confidence and safety to those facing the mob. If a lethal threat is presented, the DM team in an overwatch position (armed with a standard infantry rifle that is mounted with a high-powered scope) can scan the crowd, identify agitators and riot leaders for apprehension, or fire lethal fire if so ordered or warranted. It is also ideally suited for flank and countersniper operations.

6-29. The DM team sets the security overwatch and provides real-time intelligence and reporting that is vital to mission success. It provides coverage during the entire approach to the crowd, and its ability to select positions and provide cover fire (if and when warranted) is critical to the safety of the control force. Successful execution requires training, practice, and rehearsal with the control force formation.

6-30. The DM team is organized from a standard three- or four-man team. Each individual on the team is equipped with a standard-issue weapon. One individual has an infantry rifle with a scope, another has a set of binoculars, and two other individuals provide security for the team. Each team is equipped with a radio for communication with the HQ element of the control force formation.

6-31. The teams must have a visual advantage over the crowd to provide lethal protection to the formation. This is best accomplished by placing them on nearby rooftops, in the upper floors of buildings, and on hilltops. They must have the optical equipment to identify mob leaders, instigators, and/or individuals with weapons. Team integrity must be maintained. In organizations with three-man teams, one individual becomes the marksman, another the observer, and another the security. When teams have a 4th member, that individual becomes the video camera operator. This will help identify individuals for later apprehension and provides a record of events.

6-32. Two teams should be deployed to cover the flanks of the formation. Each team should be deployed to a position that is in front of the formation and in an overwatch position. As the formation moves forward, the teams leapfrog forward from the formation.

6-33. The reserve forces, located at the rally point, can provide another lethal force of some size that can be brought forward, as needed. Each soldier in the control force formation should have a sidearm (9-millimeter pistol) or an M16 or M4 rifle. The rifle is slung diagonally across his back with the rifle butt over his left shoulder and the muzzle below his right hip. Lethal munitions are in a separate ammunition pouch that is isolated from NL munitions. This is to prevent the mixing of NL and lethal ammunition.

6-34. Once the lethal overwatch teams are deployed, they are under the OPCON of the formation commander. The lethal overwatch element must have direct and secure communications with the formation commander. The number of teams deployed is based on METT-TC. Teams will leapfrog to keep ahead of the advancing control force formation with one team in the overwatch position at all times.

RESERVE FORCES

6-35. A large reserve of soldiers should be maintained during civil disturbance operations. Knowing that a large reserve force is available provides confidence and safety within the control force and helps prevent them from overreacting to provocative acts with disorderly and criminal elements in the crowd. When determining the number of reserve forces required, it should be resolved in favor of a large number. Lethal overwatch teams, forces to augment the control force, and apprehension teams come from the reserve forces.

6-36. Apprehension teams are a must when it is likely that a large number of people will be apprehended. These teams provide an organized response to what could be a chaotic situation. The teams can be organized at the team or squad level, depending on the number of apprehensions expected. These teams usually operate behind the base element. During peaceful demonstrations, these teams can be sent forward to remove demonstrators who refuse to disperse.

6-37. Each team consists of an apprehension, movement, and security element (added if the control force cannot provide security); a recorder; and a video or still photographer. The apprehension element is an officer or NCO and a civilian police officer (CONUS or OCONUS) or US marshal. It decides who will be apprehended, informs the subject of the reasons, supervises the handling of the subject, and ensures that the apprehension is properly conducted and documented. The apprehension team is responsible for the overall apprehension process, limiting the number of individuals appearing in court to document the apprehension. The movement element helps the apprehension element handle the subject. They move, restrain, and search the individuals being apprehended under the supervision of the apprehension element. The security element prevents the crowd from interfering with the mission. The recorder documents the apprehension on *DD Form 2708*. When possible, the video and/or still photographer recorder photographs each subject with the apprehension element. This provides the apprehension element with the necessary documentation for use in court. One individual may need to conduct many of the duties of an apprehension team based on METT-TC.

6-38. The video and/or still photographer record the situation before, during, and after the apprehensions. These images may also be used as pictorial documentation for use in court. It is possible that the presence of video and still cameras actively recording the situation can cause crowd members to be less prone to unruly or criminal acts.

6-39. The behavior of the selected individual to be apprehended is observed to decide if the subject is violent, cooperative, or passive. Unless there is good reason to believe otherwise, the subject must be considered dangerous. The mood of the crowd is observed to estimate what it might do. Is it only curious onlookers or does it support the individual targeted for apprehension? Will it interfere?

6-40. When apprehending an individual, the actions must be kept as simple as the situation allows. The element of surprise can give the apprehension team an advantage over the apprehended individual. It is best not to use surprise if the individual might panic and injure someone. Select an area for the apprehension that is to the advantage of the team and not the individual. Avoid apprehensions in crowded areas as these areas provide many avenues of escape and contain supporters of the apprehended individual.

6-41. All verbal commands given to the apprehended individual must convincing authority with short, clear, and distinct commands. Voice, attitude, and deportment must convey an expectation that the apprehended individual will comply.

6-42. The individual should be searched immediately; however, circumstances (violence or sympathizers for the individual being apprehended) may dictate temporarily postponing the search until the individual can be moved to a safer area. Flex-Cufs or hand irons may be applied.

SQUAD FORMATIONS

6-43. Civil disturbance training is a collective task. Using team and squad concepts, team leaders instruct their teams and squad leaders instruct their squads. Safety is key when using force on force training for the control force formation. Soldiers must understand that serious injuries may occur if they do not follow guidelines set by the commander.

6-44. The smallest formation is a squad formation, which is used in back of the main formation to cover side streets. Squad members must know their positions regardless of which control force formation they are in, squad to company.

6-45. Training must enforce soldiers' understanding of the need to stay informed for individual protection. Individual soldiers are much more vulnerable to attack when they break ranks and chase after crowd members. When an individual soldier breaks ranks he not only puts himself in danger, but the entire formation.

6-46. When moving a squad into a crowd control formation from a column formation, the squad leader takes one or more steps to the right of the squad and faces it. He gives the command and hand-and-arm signal for the formation he wants. Then he indicates where he wants the formation located by pointing in that general direction. The A team leader commands, "Follow Me." He places the baseman at the position indicated by the squad leader. The squad forms on the baseman position.

LINE FORMATION

6-47. The command for forming a squad in a line formation is "Squad Line Formation, Move." On the command of execution, the A team leader leads the baseman to the place indicated by the squad leader. The squad members align themselves in sequence with the baseman at normal intervals (see *Figure 6-6*). Based on the situation, the squad leader may designate a specific interval. He does this when he gives the preparatory command. If no interval is specified, the squad automatically forms using the normal interval.

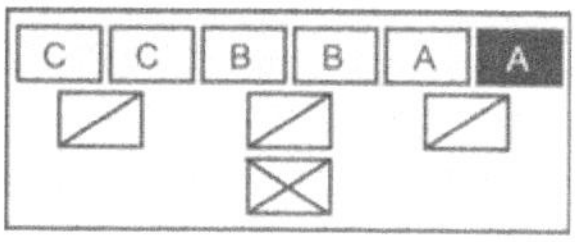

Figure 6-6. Squad Line Formation

ECHELON FORMATION

6-48. The command for having a squad form in an echelon formation is "Squad Echelon Right (or Left), Move" (see *Figure 6-7*). On the command of execution, the A team leader places the baseman at the location indicated by the squad leader. The squad members align themselves in sequence with the baseman, one pace to the side and one pace to the rear.

WEDGE FORMATION

6-49. The command for moving a squad formation into a wedge formation is "Squad Wedge, Move." On the command of execution, the A team leader moves the baseman to the place indicated by the squad leader. The A team lines up to the left of the baseman, one pace to the left and one pace to the rear of each preceding man. B team members align themselves with the baseman, one pace to the right and one pace to the rear of each preceding man. C team members align themselves in the same way to the left of the baseman (see *Figure 6-8*).

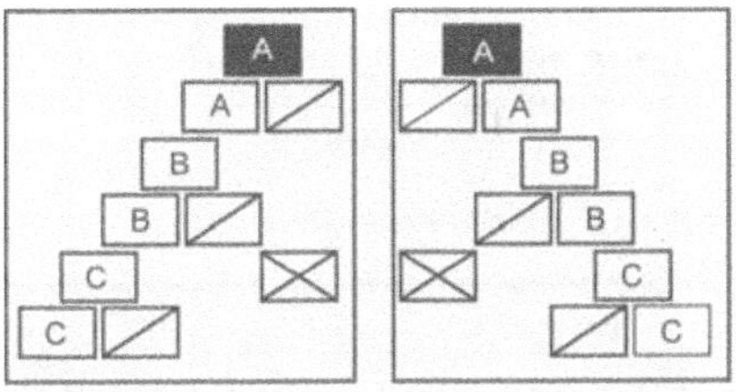

Figure 6-7. Squad Echelons Left and Right Formations

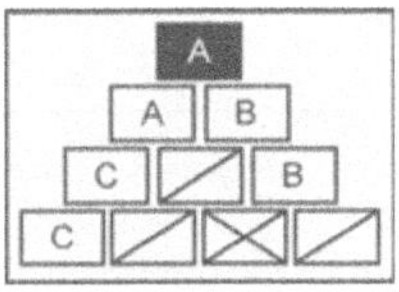

Figure 6-8. Squad Wedge Formation

ASSEMBLY

6-50. When assembling a squad from the line formation, the squad leader takes a position a sufficient distance to the rear of the squad and commands, "Squad Assemble." At the same time, he raises his right hand in the air and makes a circular motion. Squad members, who hear the command, automatically come to the port arms position. The A team leader and the baseman do an about-face movement.

6-51. The other squad members face toward the baseman. On the command of execution "Move," the squad leader points to the place where he wants the squad to assemble. The A team leader double-times to the designated spot, and the other members of the squad follow him. Team leaders fall into the file in their designated positions as they move to the AA. To assemble from a squad echelon formation, the steps are the same as a line formation.

6-52. To assemble from a squad wedge formation, the squad leader takes the same steps as for the line and echelon formations. B team members do a half left face, and the A team and C team members do a half right face. The C team pauses at the baseman position and allows the B team to clear the formation before moving to the AA.

DIAMOND AND CIRCLE FORMATIONS

6-53. The diamond and circle formations are used during extraction team operations. They are small formations used to penetrate the crowd or cover small areas. The decision on which formation to use is based on METT-TC and the ability of the squad to perform.

6-54. Commands for forming a squad in a diamond or circle formation is "Squad Diamond (or Circle), Move." On the command of execution, the A team leader moves the baseman to the place indicated by the squad leader. The A team lines up to the left of the baseman, covering the 8 to 12 o'clock positions. B team members align themselves to the right of the baseman, covering the 12 to 4 o'clock positions. C team members align themselves covering the 4 to 8 o'clock positions (see *Figure 6-9*).

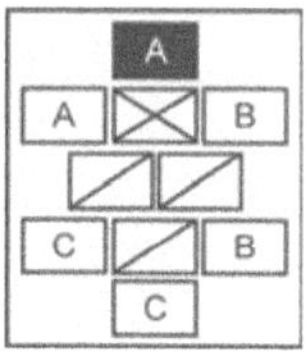

Figure 6-9. Squad Diamond or Circle Formation

6-55. To assemble from a squad diamond or circle formation, the squad leader takes the same steps as he would for a line formation. The A team leads the squad to the AA followed by the B team and then the C team. Teams fall into a column formation as they file back to the AA.

THREE-SQUAD PLATOON FORMATIONS

6-56. The platoon HQ consists of the following individuals: the platoon leader, the PSG, and a messenger or radio operator. When directed by the platoon leader, other individuals such as the selected marksman, interpreter, RCA disperser operators, and firefighters may augment the platoon HQ. In forming all control force formations from the column, the platoon leader moves to the right of the platoon and faces them when giving his commands. He gives the hand-and-arm signal and the preparatory command. On the command of execution "Move," he points to the approximate location where he wants the platoon to form. The platoon leader pauses between the preparatory command and the command of execution to allow each squad leader to issue a preparatory command to his squad.

Platoon Line Formation

6-57. The command for forming a platoon line is "Platoon on Line, Move." Immediately following the platoon leader's preparatory command, the squad leaders of the 1st and 3d squads give the command, "Follow Me." The leader of the 2d squad gives the command, "Stand Fast." On the platoon leader's command of execution, the 1st and 3d squads move forward to the point designated by the platoon leader. The 1st team member of the 1st squad is the baseman for the platoon formation. The 1st squad forms a squad line to the left of the baseman while the 3d squad forms a squad line to the right of the baseman and dresses with the 1st squad. After the 1st and 3d squads have cleared the platoon, the 2d squad leader gives the command, "Follow Me." The 2d squad splits and the A and B teams form a line to the right of the 3d squad.

The C team (with the squad leader) forms a line to the left of the 1st squad (see *Figure 6-10*).

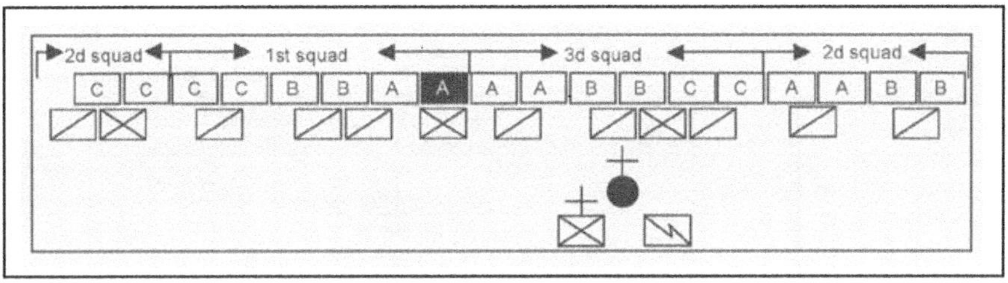

Figure 6-10. Platoon Line Formation

Assembling From a Platoon Line Formation

6-58. When assembling the platoon from the platoon line formation, the platoon leader and his HQ personnel take a position to the rear of the platoon. He then gives the command, "Platoon Assemble," while raising his right arm and making a circular motion above his head. Platoon members automatically assume the port arms position. The 1st squad leader and the baseman do an about face. The 3d squad leader moves to the head of his squad. All other members do an about-face movement toward the baseman. On the platoon leader's command of execution, the platoon leader points to where he wants the platoon to assemble. The 1st squad leader gives the command, "Follow Me," and moves his squad toward the AA. The squad moves to make a single file line to the baseman position followed by the C team of the 2d squad. As the 1st squad clears the baseman position, the 2d squad leader comes to a halt at the baseman position. The 3d squad leader gives the command, "Follow Me," and leads his squad to the AA followed by the A and B teams of the 2d squad. The A team leader of the 2d squad comes to a halt at the baseman position. Once the 3d squad has cleared the baseman position, the 2d squad leader gives the command, "Follow Me," and leads his squad to the AA. The 2d squad reassembles into their original column formation as they move to the AA. As the 1st squad comes to a halt (facing the crowd) in the AA, the 3d squad dresses with the 1st squad and leaves space for the 2d squad to return to its place in the formation. Once the platoon is formed into a column formation, HQ personnel take their position in the formation. The PSG then gives the command, "Countercolumn March," and moves the platoon to the rally point.

Platoon Line Formation With General Support

6-59. Whenever the command for establishing a crowd control formation contains the phrase "In Support" (without modification), it means that the support element is to remain in general support.

6-60. The command for forming a platoon line with the 2d squad in general support is "Platoon Line, 2d Squads in Support, Move." The 1st and 3d squads of the platoon line with general support execute the line, while the 2d squad remains in the column (see *Figure 6-11*). To assemble a platoon with a squad in general support, the procedure is the same as the platoon line, except that the 1st and 3d squads dress with the 2d squad.

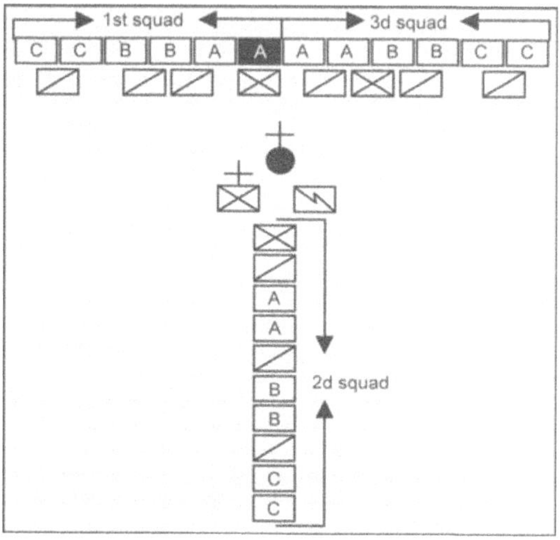

Figure 6-11. Platoon Line Formation With General Support

Platoon Line Formation With Lateral Support

6-61. The command for forming a platoon line with one squad in lateral support is "Platoon Line, 2d Squad in Lateral Support, Move." The 1st and 3d squads execute the line as before, and the 2d squad stands fast. After the line has been formed, the 2d squad leader gives the command, "Follow Me." The A and B teams move to the right flank while the squad leader and the C team move to the left flank of the formation. The 2d squad forms columns behind the last men on the line formation. If no direction is given, the 2d squad splits to cover both flanks. A squad may be used to cover one flank by giving the command, "2d Squad Right (or Left) Lateral Support" (see *Figure 6-12*).

6-62. The 2d squad may be moved from general to lateral support at any time by the platoon leader. The platoon leader commands, "2d Squad, Lateral Support, Move." To have the 2d squad join the line from either support position, the platoon leader commands, "2d Squad, Extend the Line, Move." The 2d squad leader commands, "Follow Me." On the command of execution, the squads set up individual lines, closing and dressing on the existing line. To assemble the platoon from a platoon line with lateral support, follow the same steps as for the platoon line.

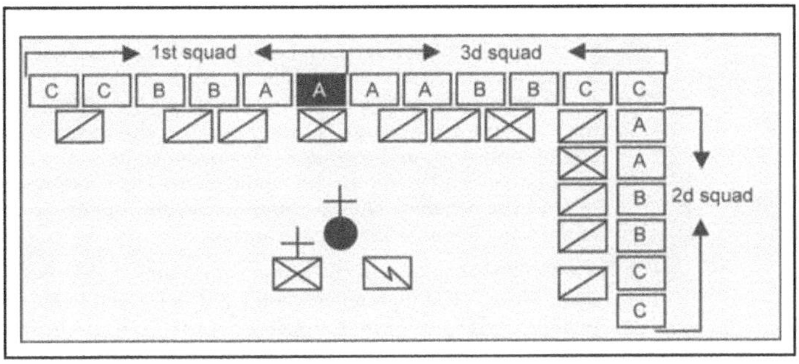

Figure 6-12. Platoon Line Formation With Lateral Support (Right)

Platoon Line Formation With Direct Support

6-63. The command for forming a platoon line with one squad in direct support is "Platoon Line, 2d Squad in Direct Support, Move." The 1st and 3d squads execute a line as before. The 2d squad executes a squad line directly to the rear and centered on the 1st line. The men in the supporting line cover the intervals between the men in the 1st line. To do this, the squad and team leaders of the 1st and 3d squads must take two steps to the rear to allow space for the 2d squad to move into place. After seeing the squad and team leaders move back from the line, the 2d squad leader commands, "Follow Me," and leads his squad into position (see *Figure 6-13*).

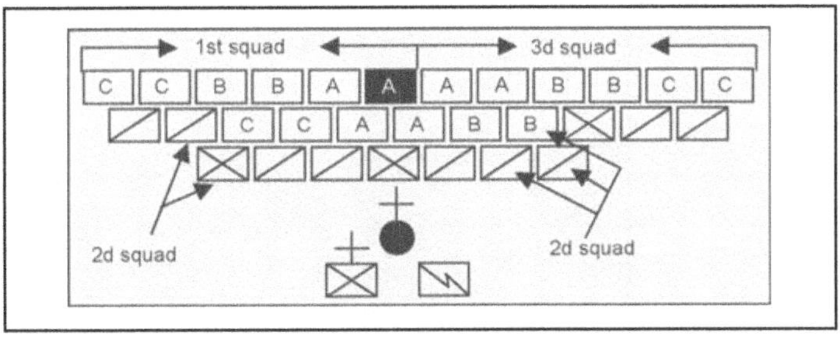

Figure 6-13. Platoon Line Formation With Direct Support

6-64. To assemble a platoon with direct support, the procedure is the same as the platoon line, except that the 2d squad moves to the AA first. To assemble the support squads from any position to general support, the platoon leader commands, "2d Squad in Support, Move." The 2d squad then returns to a column behind the line formed by the other two squads.

Platoon Echelon Right and Left Formations

6-65. The command for forming a platoon echelon right is "Platoon Echelon Right, Move." After the platoon leader's preparatory command, the squad leader of the 1st squad commands, "Follow Me." The squad leaders of the 2d and 3d squads command, "Stand Fast." On the command of execution, the 1st squad moves out and executes an echelon right at the location designated by the platoon leader. As the 1st squad clears the column, the 3d and 2d squads extend the echelon. The 2d squad performs support to this formation in the same ways it supports a line formation.

6-66. The command for forming a platoon echelon left is "Platoon Echelon Left, Move." The platoon echelon right is formed in the same manner as the echelon left (see *Figure 6-14*). *Figures 6-15; 6-16,* page 6-22; and *6-17,* page 6-23 show support formations in a left echelon.

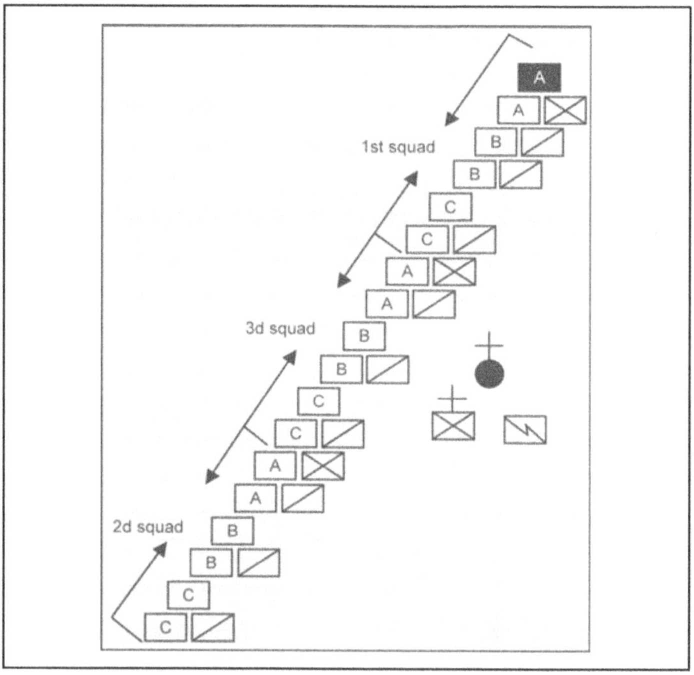

Figure 6-14. Platoon Echelon Left Formation

6-67. To assemble the platoon, the commander takes his position to the rear of the formation with the platoon HQ. On the preparatory command, "Platoon Assemble," the 1st squad leader does an about-face movement while all other

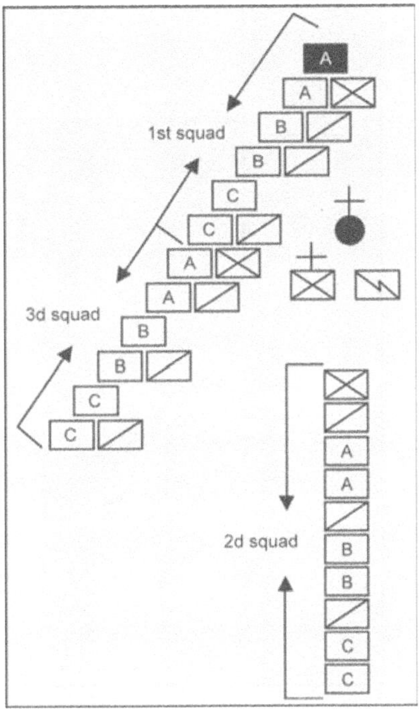

Figure 6-15. Platoon Echelon Left Formation With 2d Squad in General Support

members assume the port arms position and face the baseman. On the command, "Move," the 1st squad moves first, followed by the 3d and 2d squads. Each squad dresses with the 1st squad as it reaches the AA. The assembly of the echelon right formation is the same as the echelon left.

Platoon Wedge Formation

6-68. The command for forming the platoon wedge formation is "Platoon Wedge, Move." After the platoon leader's preparatory command, the squad leaders of the 1st and 3d squads command, "Follow Me." At the same time, the 2d squad leader commands, "Stand Fast." On the platoon leader's command of execution, the 1st and 4th squads move directly to the front. The 1st squad executes an echelon left on the baseman while the 3d squad executes an echelon right. When the last man of the 1st and 4th squads clear the 2d squad, the 2d squad leader commands, "Follow Me," and moves out to the left and right. The 2d squad splits and the A and B teams go to the right and the C

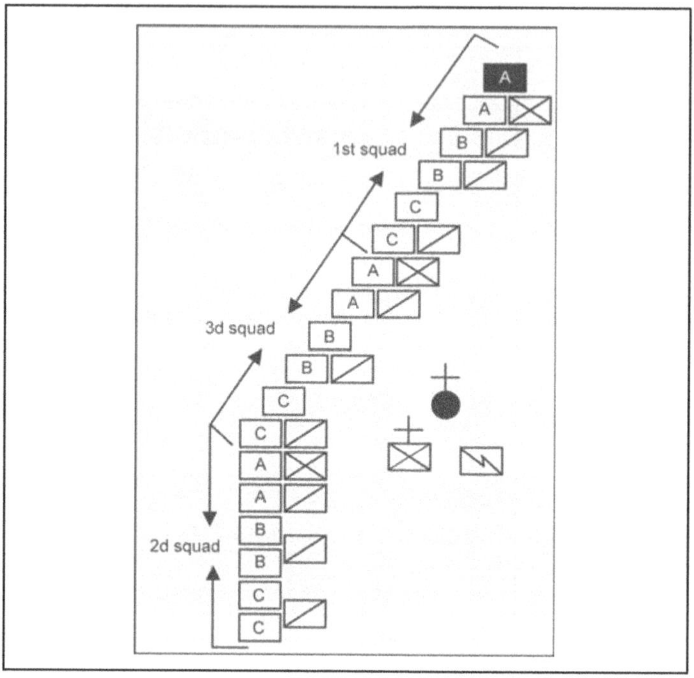

Figure 6-16. Platoon Echelon Left Formation With 2d Squad in Lateral Support

team goes to the left to extend the formation (see *Figure 6-18*). Assembly of the wedge formation is the same as the platoon line.

Platoon Wedge Formation With General Support

6-69. The command for forming a platoon wedge formation with support is "Platoon Wedge, 2d Squad in Support, Move." The 1st and 3d squads execute the wedge while the 2d squad remains in a column formation (see *Figure 6-19,* page 6-24).

Platoon Wedge Formation With Lateral Support

6-70. The command for forming a platoon wedge formation with lateral support is "Platoon Wedge, 2d Squad in Lateral Support, Move." The 1st and 3d squads execute the wedge while the 2d squad stands fast. After the 1st and 3d squads have cleared the column, the 2d squad leader commands, "Follow Me." If no direction is given, the 2d squad splits to cover both flanks. A squad

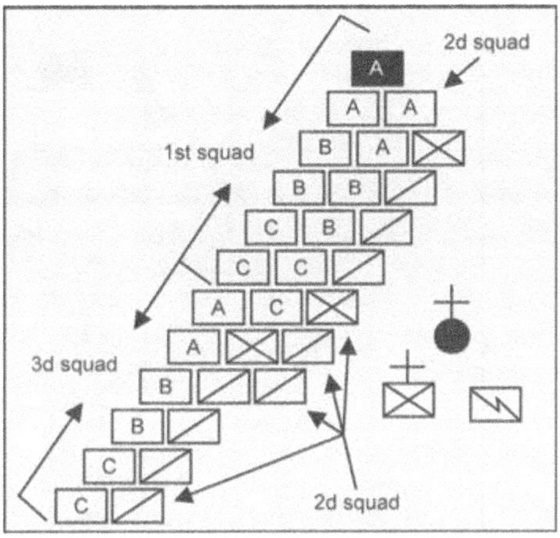

Figure 6-17. Platoon Echelon Left With 2d Squad in Direct Support

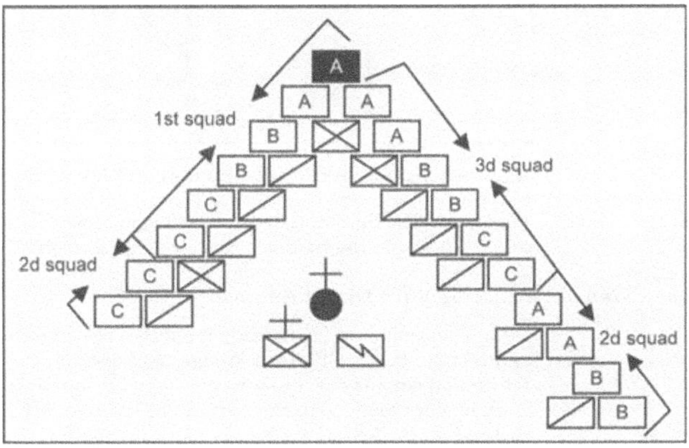

Figure 6-18. Platoon Wedge Formation

may be used to cover one flank by commanding, "2d Squad Right (or Left) Lateral Support" (see *Figures 6-20* and *6-21,* page 6-25).

6-71. To move the 2d squad from general support to lateral support, the platoon leader commands, "2d Squad, Lateral Support, Move." To have the 2d squad join the wedge from general or lateral support, the platoon leader

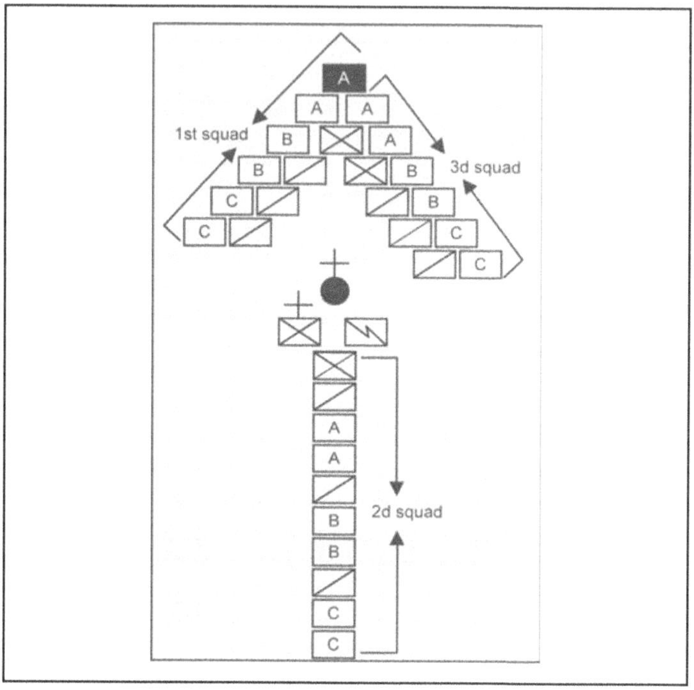

Figure 6-19. Platoon Wedge Formation With 2d Squad in General Support

commands, "2d Squad, Extend the Wedge, Move." To extend the wedge, the 2d squad splits and the teams move to the left and right, respectively.

Platoon Wedge Formation With Direct Support

6-72. The command for forming a platoon wedge formation with the 2d squad in direct support is "Platoon Wedge, 2d Squad in Direct Support, Move." The 1st and 3d squads execute a wedge. The 2d squad waits for the squad and/or team leaders to move to the rear, as in the line formation, before executing a wedge directly behind and centered on the main formation. The individuals in the supporting wedge cover the intervals between individuals in the leading wedge (see *Figure 6-22,* page 6-26).

Platoon Diamond and Circle Formations

6-73. These formations are used in open areas or where the crowd may be able to envelop the formation. The diamond and circle formations are primarily defensive in nature and difficult to maneuver. The decision about which formation to use is based on the ability of the platoon to perform them.

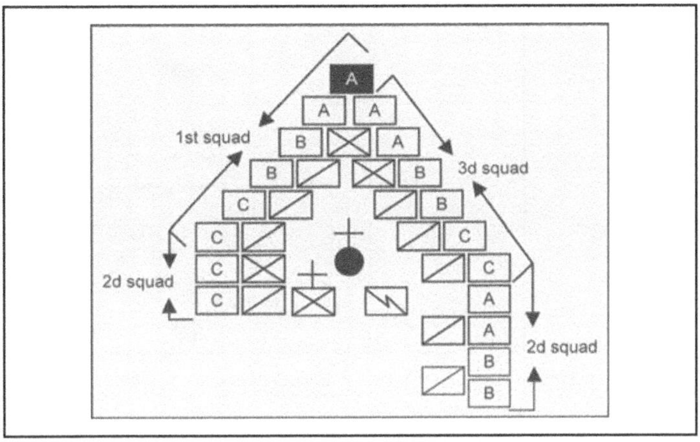

Figure 6-20. Platoon Wedge Formation With 2d Squad in Lateral Support

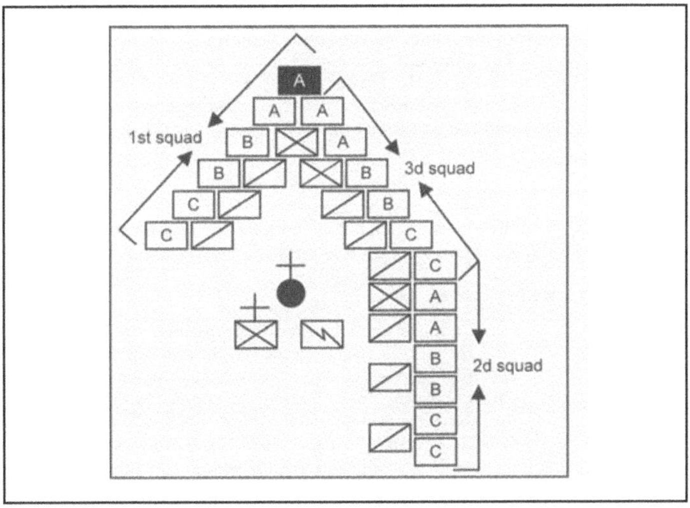

Figure 6-21. Platoon Wedge Formation With 2d Squad in Right Lateral Support

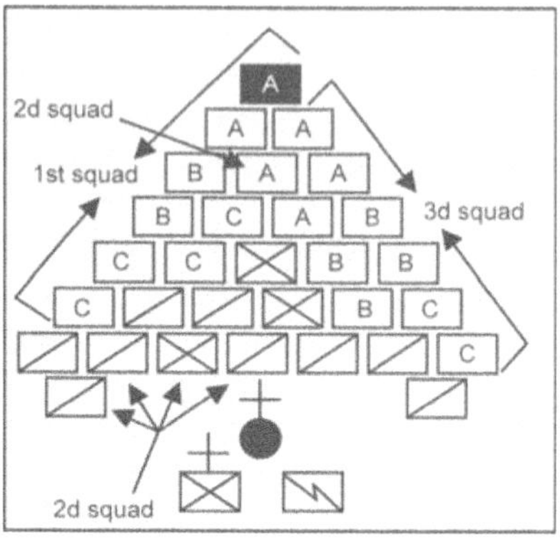

Figure 6-22. Platoon Wedge Formation With 2d Squad in Direct Support

The command for having a platoon form in a diamond or circle formation is "Platoon Diamond (or Circle), Move." On the command of execution, the 1st squad leader moves the baseman to the place indicated by the squad leader. The baseman becomes the 12 o'clock position while the 1st squad lines up to the left of the baseman covering the 8 to 12 o'clock positions. The 3d squad aligns itself to the right of the baseman, covering the 12 to 4 o'clock positions and the 2d squad aligns itself, covering the 4 to 8 o'clock positions (see *Figures 6-23* and *6-24*).

6-74. To assemble from a platoon diamond or circle formation, the HQ element takes the same steps as that of a line formation. The 1st squad leads the platoon back to the AA followed by the 2d and 3d squads. Squads fall into a column formation as they file back to the AA.

RELIEF ELEMENTS IN PLACE

6-75. During crowd control operations, the platoon leader may rotate his squads to give them a rest. He is not limited to using only those squads specified in the preceding examples as the lead elements of the formations. He can replace the base elements by commanding direct support to the 1st squad and then directing the 1st squad into general support. The support element (the 2d squad) then passes through the 1st squad. Once the 2d squad is formed directly behind the 1st squad, the 1st squad leader commands, "Right Face." The members of the 1st squad do a right-face movement and the 2d squad steps quickly though the 1st squad and forms the base formation. The 1st squad leader then commands, "Follow Me," and leads his squad to the general support position. This procedure is often necessary when protective masks are used and chemical irritants employed. The 1st squad then replaces the 3d squad in the same manner. Platoons participating in company operations are relieved in the same manner.

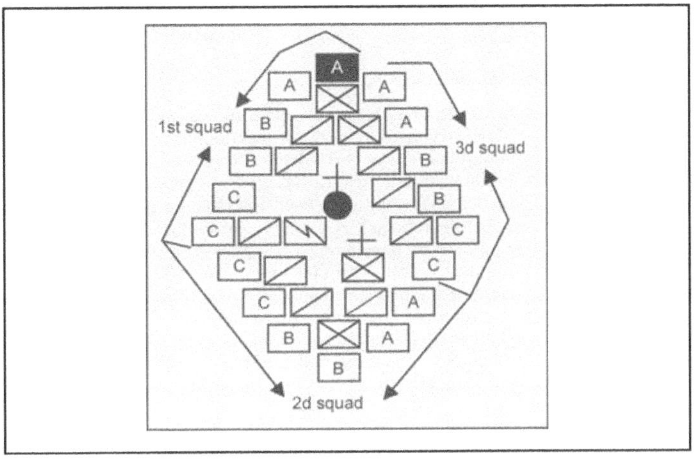

Figure 6-23. Platoon Diamond Formation

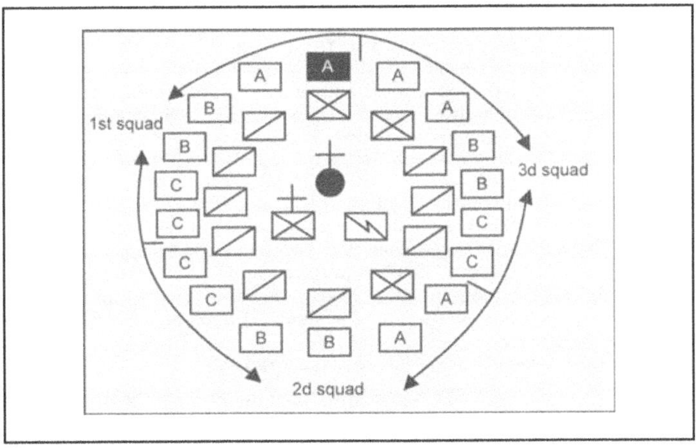

Figure 6-24. Platoon Circle Formation

FOUR-SQUAD PLATOON FORMATIONS

6-76. When forming control force formations from the column, the platoon leader moves to the right of the platoon and faces it when giving his commands. He gives the hand-and-arm signal and the preparatory command. On the command of execution, "Move," he points to the approximate location where he wants the platoon to form. The platoon leader pauses between the preparatory command and the command of execution to allow each squad leader to issue a preparatory command to his squad.

Platoon Line Formation

6-77. The command for forming a platoon line formation is "Platoon Line, Move." Immediately following the platoon leader's preparatory command, the squad leaders of the 1st and 4th squads command, "Follow Me." The squad leaders of the 2d and 3d squads command, "Stand Fast." On the platoon leader's command of execution, the 1st and 4th squads move forward to the point designated by the platoon leader. The 1st A team member of the 1st squad is the baseman for the platoon formation. The 1st squad forms a squad line to the left of the baseman and the 4th squad forms a squad line to the right of the baseman with all lines dressed to the right. After the 1st and 4th squads have cleared the platoon, the 2d and 3d squad leaders command, "Follow Me." The 2d squad forms a squad line to the left of the 1st squad. The 3d squad forms a squad line to the right of the 4th squad. All squads dress with the 4th squad (see *Figure 6-25*).

Platoon Line Formation Assembly

6-78. When assembling from the platoon line formation, the platoon leader and his HQ personnel take a position to the rear of the platoon. He then gives the command, "Platoon Assemble," while raising his right arm and making a circular motion above his head. Platoon members automatically assume the port arms position. The 1st squad leader and the baseman do an about-face movement. All other platoon members do an about-face movement toward the baseman. On the platoon leader's command of execution, he points to where he wants the platoon to assemble. He commands, "Follow Me," and moves his squad toward the AA. The 1st squad files to the baseman position, followed by the 2d squad. As the 2d squad clears the baseman position, the 4th squad leader then commands, "Follow Me," and leads his squad to the AA, followed by the 3d squad. As the 1st squad comes to a halt facing the crowd in the AA, the 2d squad assumes its position to the right of the 1st squad. As the 4th squad assembles, it dresses on the 1st squad, leaving space for the 3d squad to return to its place in the formation. Once the platoon is formed into a column formation, HQ personnel take their position in the formation. The PSG then commands, "Countercolumn, March," and moves the platoon to the rally point.

Platoon Line Formation With General Support

6-79. Whenever the command for establishing a crowd control formation contains the phrase "In Support Without Modification," it means that the support element is to remain in general support. The command for forming a platoon line with two squads in general support is "Platoon Line, 2d and 3d Squads in Support, Move." The 1st and 4th squads execute the line, while the 2d and 3d squads remain in the column (see *Figure 6-26,* page 6-30). To assemble a platoon with two squads in general support, the procedure is the same as that of the platoon line, except that the 1st and 4th squads dress on with the 2d and 3d squads.

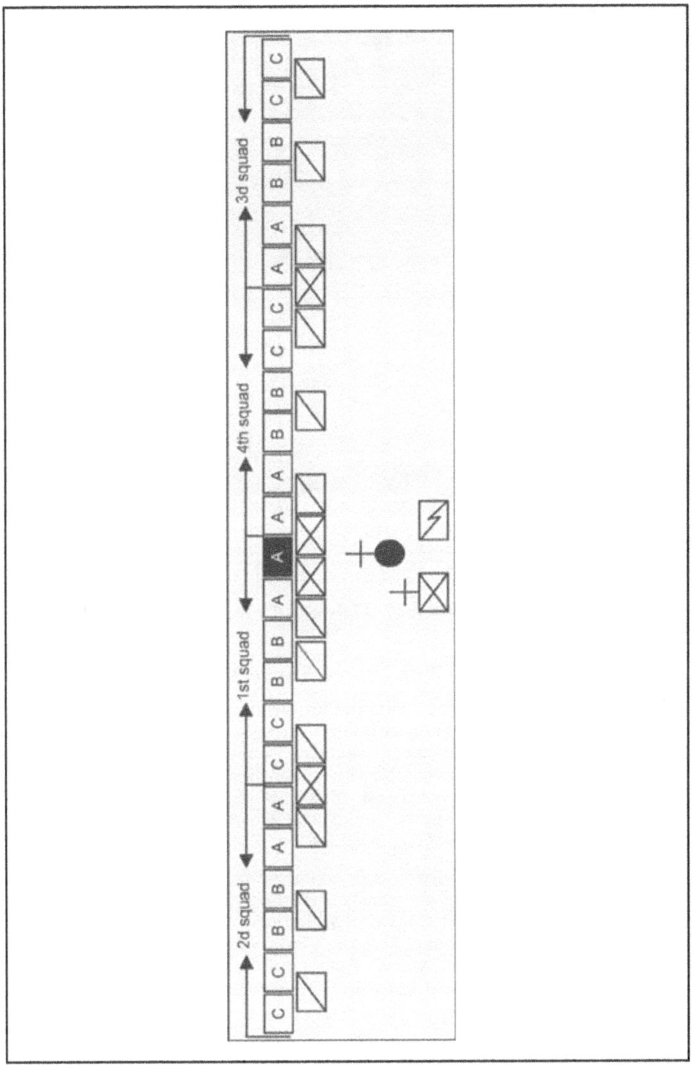

Figure 6-25. Platoon Line Formation With Four Squads

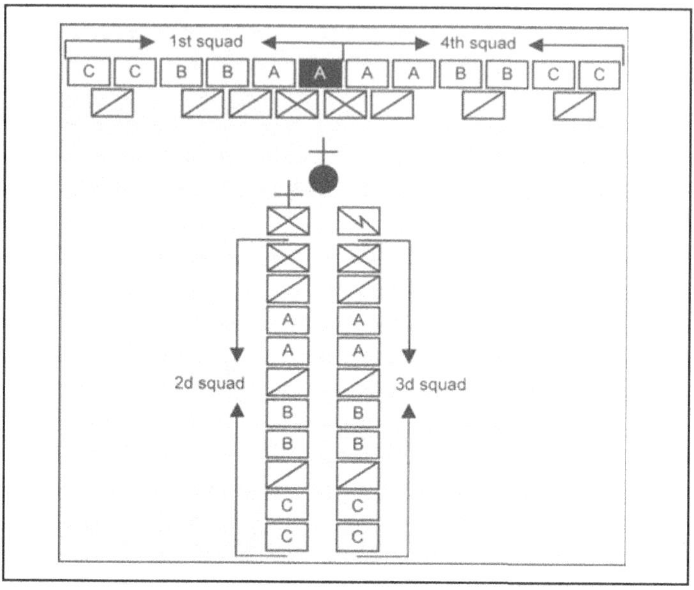

Figure 6-26. Platoon Line Formation With d and 3d Squads in General Support

Platoon Line Formation With Lateral Support

6-80. The command for forming a platoon line formation with two squads in lateral support is "Platoon Line, 2d and 3d Squads in Lateral Support, Move." The 4th and 1st squads execute the line as before, and the 2d and 3d squads stand fast. After the line has been formed, the squad leaders of the 2d and 3d squads command, "Follow Me." The 2d squad forms a column behind the last man of the 1st squad. The 3d squad forms a column behind the last man of the 4th squad (see *Figure 6-27*).

6-81. Squads may be moved from general support to lateral support at any time by the platoon leader. He commands, "2d and 3d Squads, Lateral Support, Move." To have the 2d and 3d squads join the line from either support position, the platoon leader commands, "2d and 3d Squads, Extend the Line, Move." The 2d and 3d squad leaders command, "Follow Me." On the command of execution, the squads set up individual lines, closing and dressing with the existing line. To assemble the platoon from a platoon line with lateral support, the procedure is the same as that of the platoon line, except lateral support squads face forward until they move to the AA.

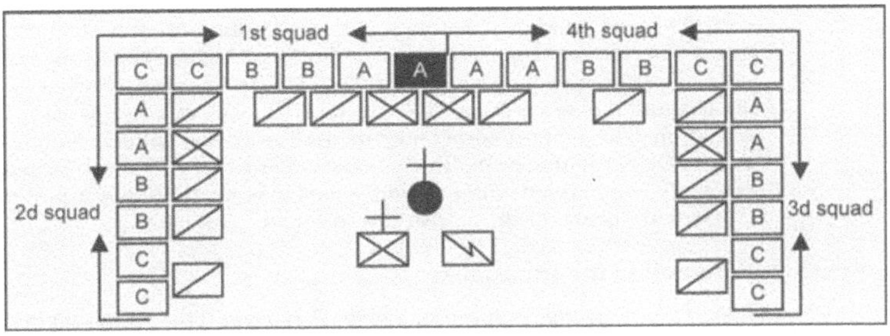

Figure 6-27. Platoon Line Formation With 2d and 3d Squads in Lateral Support

Platoon Line Formation With Direct Support

6-82. The command for forming a platoon line with two squads in direct support is "Platoon Line, 2d and 3d Squads in Direct Support, Move." The 1st and 4th squads execute a line as before. The 2d and 3d squads execute a 2d line directly to the rear and centered on the 1st line. The members in the direct support line cover the intervals between members in the baseline. To do this, the squad and team leaders of the 1st and 4th squads must take two steps to the rear to allow space for the 2d and 3d squads to move into place. After seeing the squad and team leaders move back from the line, 2d and 3d squad leaders command, "Follow Me," and lead their squads into position (see *Figure 6-28*).

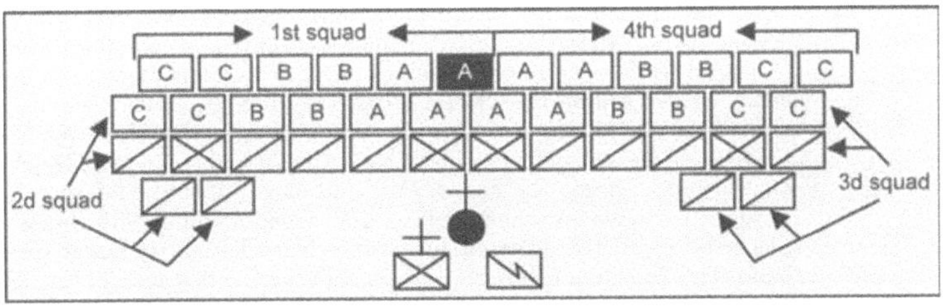

Figure 6-28. Platoon Line Formation With 2d and 3d Squads in Direct Support

6-83. To assemble the support squads from any position to general support, the platoon leader commands, "2d and 3d Squads in Support, Move." The 2d and 3d squads then return to a column behind the line formed by the 1st and 4th squads. To assemble the entire platoon, the procedure is the same as that of a platoon line, except that the 2d squad moves to the AA first, followed by the 3d, 1st, and 4th squads.

Platoon Line Formation With One Support Squad

6-84. The command for forming a platoon line formation with one squad in support is "Platoon Line, 3d Squad in Support, Move." The 1st squad moves out and executes a squad line. The 2d and 4th squads form a line on the side of the left and rights sides of the 1st squad. The 3d squad remains in column formation. The support squad may be used in lateral support on one or both sides of the formation or in direct support of any segment of the formation. To assemble the platoon, the procedure is the same as that of a platoon line, except that all squads dress with the 3d squad.

Platoon Echelon Right and Left Formations

6-85. The command for forming a platoon echelon right formation is "Platoon Echelon Right, Move." After the platoon leader's preparatory command, the squad leader of the 1st squad commands, "Follow Me." The squad leaders of the 2d, 3d, and 4th squads command, "Stand Fast." On the command of execution, the 1st squad moves out and executes an echelon right at the location designated by the platoon leader. As each squad clears the column, the next successive squad moves out and extends the echelon that has been formed. To assemble the platoon, the procedure is the same as that of a platoon line, except that all squads file back in order.

6-86. The command for forming a platoon echelon left formation is "Platoon Echelon Left, Move." The platoon echelon left is formed in the same manner as the echelon right, but in reverse order (see *Figure 6-29*). To assemble the platoon, the procedure is the same as that of an echelon right. See *Figures 6-30,* page 6-34*;* *6-31,* page 6-35*;* and *6-32,* page 6-36, for examples of echelon formations with support.

Platoon Wedge Formation

6-87. The command for forming the platoon wedge formation is "Platoon Wedge, Move." After the platoon leader's preparatory command, the 1st and 4th squad leaders command, "Follow Me." At the same time, the 2d and 3d squad leaders command, "Stand Fast." On the platoon leader's command of execution, the 1st and 4th squads move directly to the front. When the last members of the 1st and 4th squads clear the 2d and 3d squads, the 2d and 3d squad leaders command, "Follow Me," and move out to the left and right. The 1st squad executes an echelon left on the baseman and the 4th squad executes an echelon right. The 2d squad forms an echelon left on the last man of the 1st squad. The 3d squad forms an echelon right on the last man of the 4th squad. This completes the wedge formation (see *Figure 6-33,* page 6-36). To assemble the platoon, the procedure is the same as that of a platoon line.

Platoon Wedge Formation With General Support

6-88. The command for forming a platoon wedge formation with two squads in support is "Platoon Wedge, 2d and 3d Squads in Support, Move." The 1st and 4th squads execute the wedge while the 2d and 3d squads remain in column formation (see *Figure 6-34,* page 6-37).

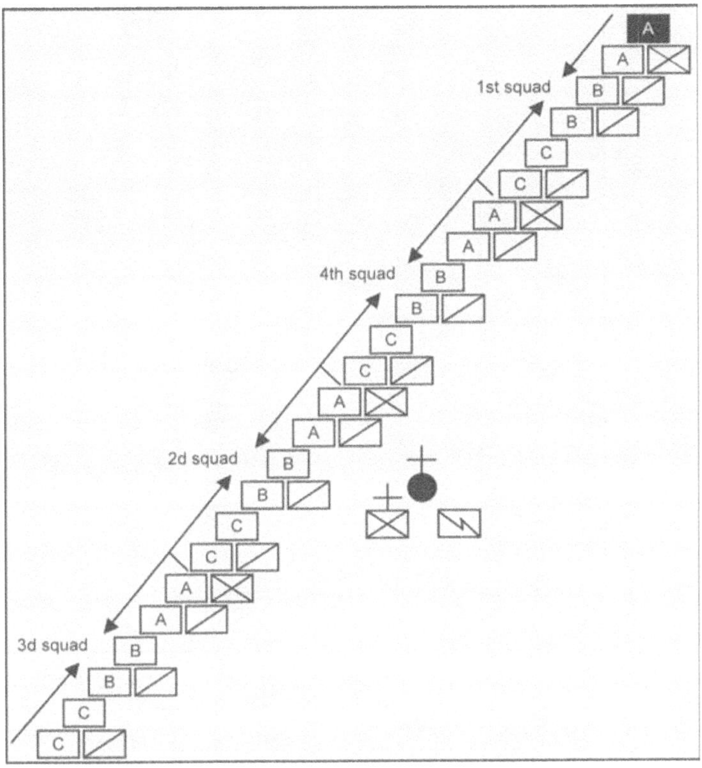

Figure 6-29. Platoon Echelon Left Formation With Four Squads

Platoon Wedge Formation With Lateral Support

6-89. The command for forming a platoon wedge formation with lateral support is "Platoon Wedge, 2d and 3d Squads in Lateral Support, Move." The 1st and 4th squads execute the wedge while the 2d and 3d squads stand fast. After the 1st and 4th squads have cleared the column, the 2d and 3d squad leaders command, "Follow Me," and move out to the left and right. The 2d squad forms a column formation behind the last man of the 1st squad. The 3d squad forms a column formation behind the last man of the 4th squad (see *Figure 6-35*, page 6-38).

6-90. To move the 2d and 3d squads from general to lateral support, the platoon leader commands, "2d and 3d Squads, Lateral Support, Move." To have the 2d and 3d squads join the wedge from general or lateral support, the

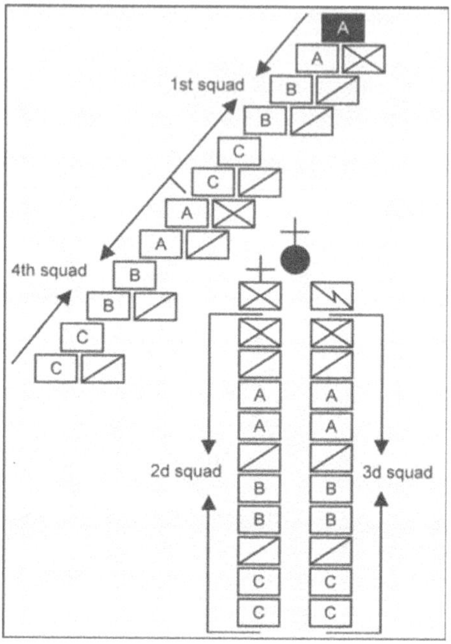

Figure 6-30. Platoon Echelon Left Formation With 2d and 3d Squads in General Support

platoon leader commands, "2d and 3d Squads, Extend the Wedge, Move." The 2d and 3d squad leaders command, "Follow Me," and move out to the left and right to extend the wedge on the 1st and 4th squads.

Platoon Wedge Formation With Direct Support

6-91. The command for forming a platoon wedge with two squads in direct support is "Platoon Wedge, 2d and 3d Squads in Direct Support, Move." The 1st and 4th squads execute a wedge. The 2d and 3d squads execute a wedge and close in on the leading wedge. The members in the supporting wedge cover the intervals between members in the leading wedge. To do this, the squad and team leaders of the 1st and 4th squads must take two steps to the rear to allow space for the 2d and 3d squads to move into place. After seeing the squad and team leaders move back from the line, the 2d and 3d squad leaders command, "Follow Me," and lead their squads into position (See *Figure 6-36,* page 6-38). To move the 2d and 3d squads back to general support, the platoon leader commands, "2d and 3d Squads Support, Move." Then the 2d and 3d squads move to their general support positions. To

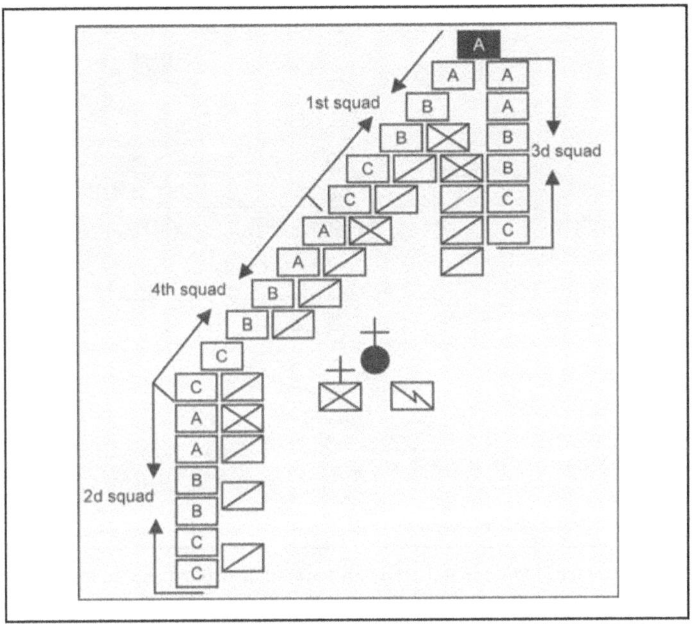

Figure 6-31. Platoon Echelon Left Formation With 2d Squad in Lateral Support

assemble a platoon from a platoon wedge with direct support, the procedure is the same as that of a platoon wedge, except that the 2d squad assembles first, followed by the 3d, 1st, and 4th squads.

Platoon Wedge Formation With One Support Squad

6-92. The command for forming a platoon wedge formation with one squad in support is "Platoon Wedge, 3d Squad in Support, Move." The 1st squad moves out and executes a squad wedge. The 2d and 4th squads form echelons left and right on the 1st squad. The 3d squad remains in the column formation.

Platoon Diamond and Circle Formations

6-93. The platoon diamond and circle formations are used in open areas or where the crowd may be able to envelop the formation. They are primarily defensive in nature and hard to maneuver. The decision concerning which formation to use is based on the ability of the platoon to perform them. The command for having a platoon form in a diamond or circle formation is "Platoon Diamond (or Circle), Move." On the command of execution, the 1st squad leader moves the baseman to the location indicated by the squad leader. The baseman becomes the 12 o'clock position. The 1st squad lines up to the

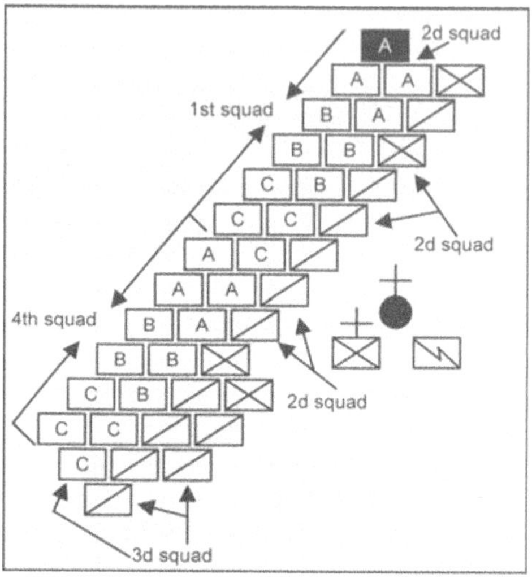

Figure 6-32. Platoon Echelon Left Formation With 2d and 3d Squads in Direct Support

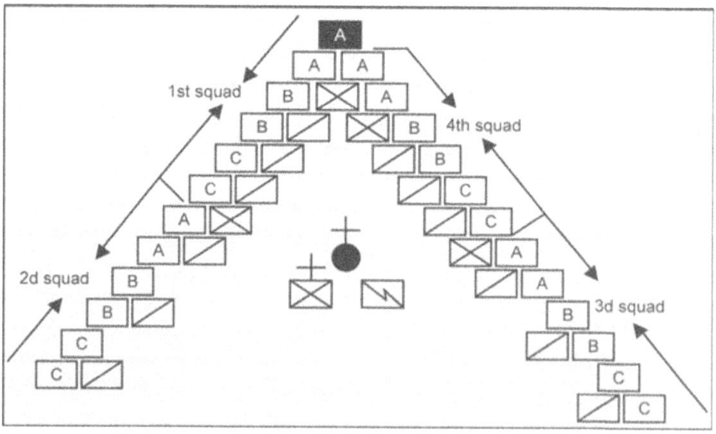

Figure 6-33. Platoon Wedge Formation With Four Squads

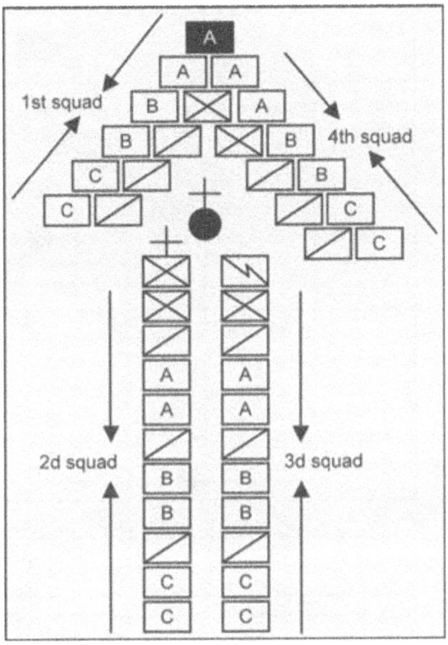

Figure 6-34. Platoon Wedge Formation With 2d and 3d Squads in General Support

left of the baseman covering the 9 to 12 o'clock positions. The 4th squad aligns themselves to the right of the baseman covering the 12 to 3 o'clock positions. The 3d squad aligns itself covering the 3 to 6 o'clock positions. The 2d squad aligns itself covering the 6 to 9 o'clock positions.

6-94. To assemble a platoon from a platoon diamond or circle formation, the HQ element uses the same procedures as that of a line formation. The 1st squad leads the platoon back to the AA followed by the 2d, 3d, and 4th squads. The squads fall into a column formation as they file back to the AA.

COMPANY FORMATIONS

6-95. When a company commander orders his company into control force formations from a column, he moves to the left near the head of the column so he can be seen by the platoon leaders. He then faces the company. As he gives his preparatory command, he points to where he wants the formation. If he does not indicate a location, the company forms immediately in front of the lead platoon.

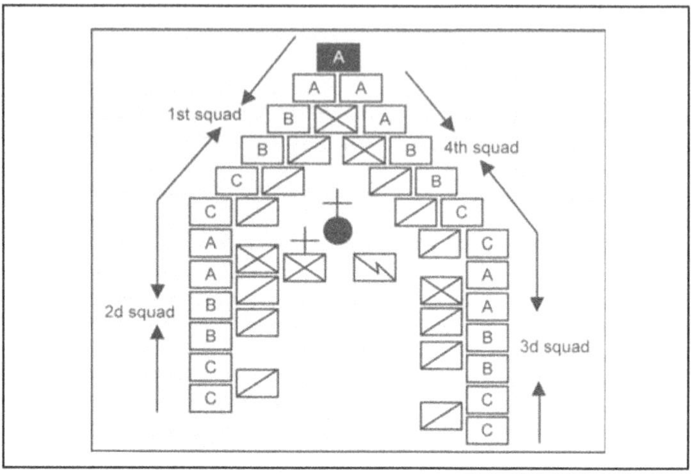

Figure 6-35. Platoon Wedge Formation With 2d and 3d Squads in Lateral Support

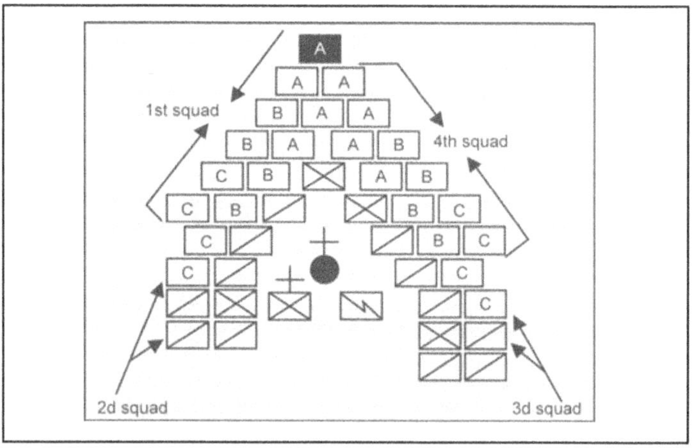

Figure 6-36. Platoon Wedge Formation With 2d and 3d Squads in Direct Support

Relieving Elements in Place

6-96. During crowd control operations, the platoon leader may rotate his squads to give them a rest. See *paragraph 6-76,* page 6-27.

Company Line in Depth Formation

6-97. The command for forming a company line in depth formation (see *Figure 6-37*) is "Company Line in Depth, Move." After the company commander's preparatory command, each platoon leader gives the command, "Platoon Line." The squad leaders follow with their commands to their squads to form the platoon line. On the command of execution, each platoon establishes a platoon line immediately to its front.

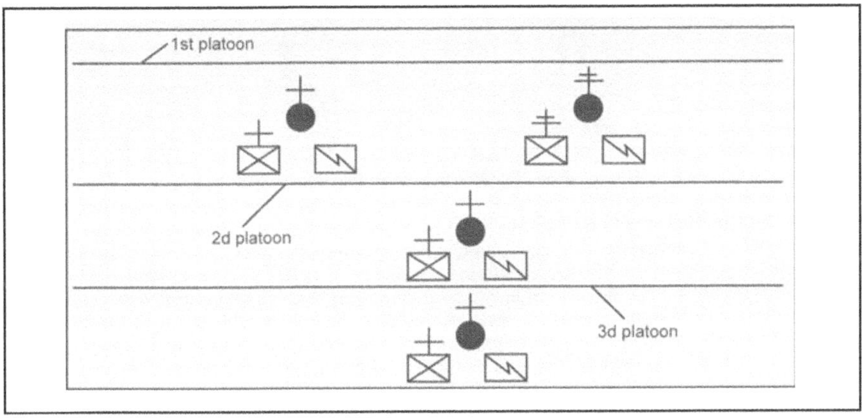

Figure 6-37. Company Line in Depth Formation

6-98. If the commander wants a more formidable formation, he commands, "2d Platoon, Direct Support, Move." The squad and team leaders from the 1st platoon take two steps backward. The 2d platoon moves forward and covers the intervals between the members of the lead platoon. The platoon leader of the 3d platoon then moves his platoon forward to occupy the position formerly held by the 2d platoon.

6-99. If the 3d platoon is called on to support the company line, it moves forward and takes up a position directly behind the 2d platoon in the same manner. Platoon leaders and PSGs help each other in controlling the company.

Company Line in Mass Formation

6-100. The command for forming a company line in mass formation is "Company Line in Mass, Move" (see *Figure 6-38,* page 6-40). The company line in mass is the same as a company line with both platoons in direct support. Instead of forming it in stages, the company commander indicates his desire for this formation in his initial command. The platoons each form a line, and the 2d and 3d platoons close on the 1st platoon without further command.

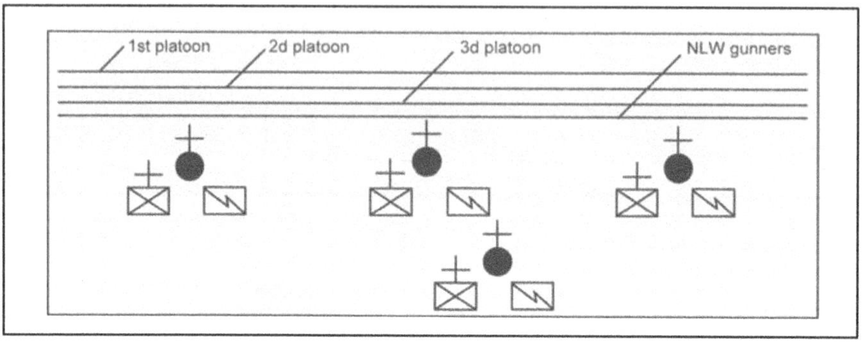

Figure 6-38. Company Line in Mass Formation

Company Line Formation With Support

6-101. In company formations, the 1st platoon usually forms the lead element and the 2d and 3d platoons provide support (see *Figures 6-39; 6-40;* and *6-41*). The support platoons can be employed in the same manner as the support squads in platoon formations. One platoon can be relieved from the lead element by another platoon while in formation by having a support platoon pass through the lead platoon.

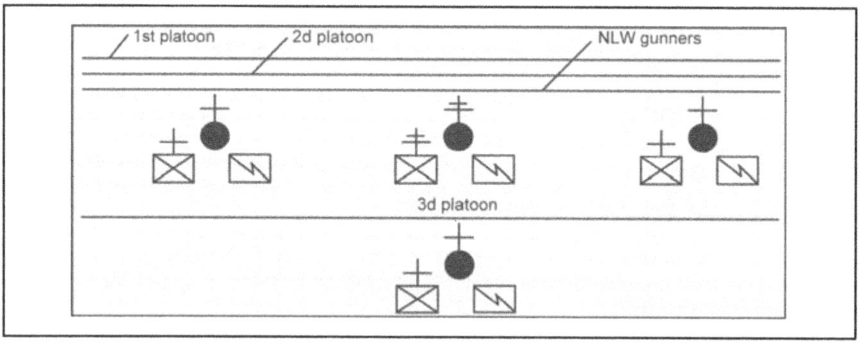

Figure 6-39. Company Line in Mass Formation With One Platoon in Depth

Company Echelons and Wedge Formations

6-102. Company echelons (see *Figures 6-42,* page 6-42*; 6-43,* page 6-43*; 6-44,* page 6-43*;* and *6-45,* page 6-44) and company wedge formations (see *Figures 6-46,* page 6-45; *6-47,* page 6-46; *6-48,* page 6-46; and *6-49,* page 6-47) are formed in the same manner and with the same variations as the company line formation. A company assembles from a crowd control formation in the same manner as a platoon or a squad does. However, due to the size of company

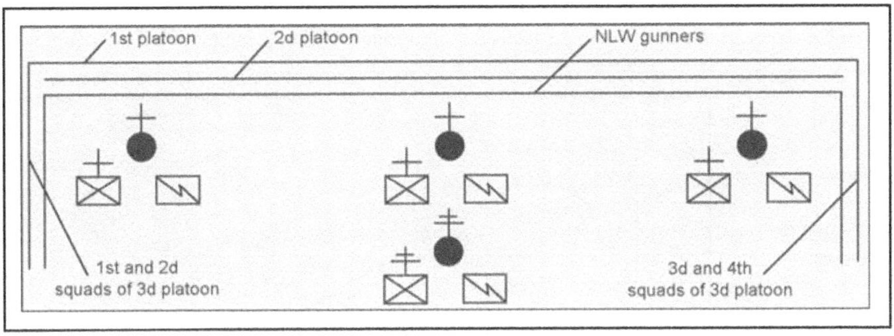

Figure 6-40. Company Line in Mass Formation With One Platoon in Lateral Support

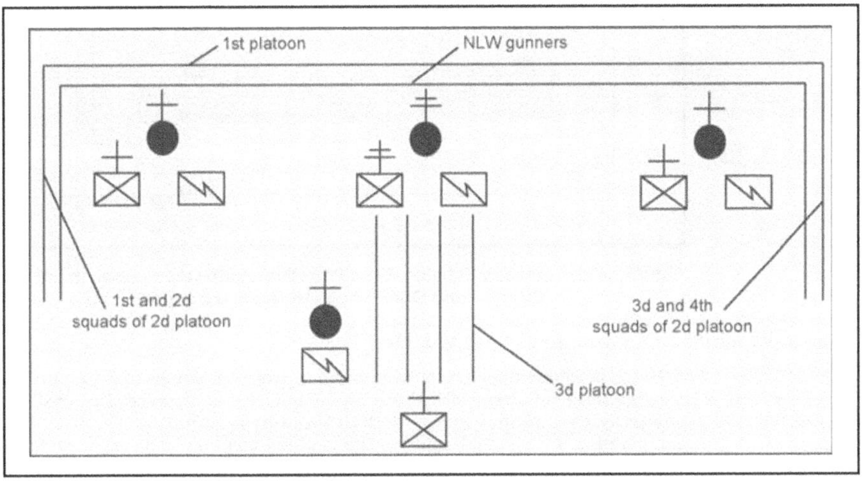

Figure 6-41. Company Line, One Platoon in Lateral Support and One Platoon in General Support

formations, the commander must consider the area that the company is operating in when he assembles the company. If he follows the guidelines of platoon assembly, he will have no difficulty. When assembling the company, the commander takes a position to the rear of the formation and commands, "Company Assemble." The platoon leaders take charge of their respective platoons. Usually, the 3d platoon assembles first, just behind the formation. After the 3d platoon is assembled, the 2d platoon assembles in front of the 3d platoon and the 1st platoon assembles in front of the 2d. All platoons assemble facing the commander. The 2d and 3d platoons then dress with the 1st platoon. As each platoon is assembled, they are ordered to port arms by the platoon leader to await further orders from the commander. If four platoons are used, the 4th platoon assembles first.

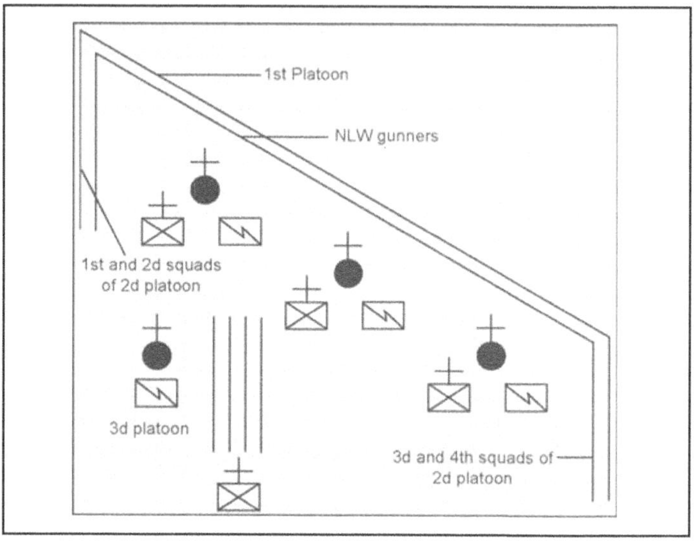

Figure 6-42. Company Echelon Right Formation With One Platoon in Lateral Support and One Platoon in General Support

Company and Platoon Formations With Vehicles

6-103. The commands for vehicles and foot troops are the same as those for foot troops alone. Because of the additional noise and distances involved, the company commander gives the hand-and-arm signal for the line. The command and the signal apply to the lead platoon only. The other platoons remain in general support until further directed (see *Figures 6-50,* page 6-48; *6-51,* page 6-49; *6-52,* page 50; *6-53,* page 6-50; *6-54,* page 6-51; *6-55,* page 6-52; and *6-56,* page 6-53).

6-104. The motor section moves out with vehicle number two moving to the right. Vehicle number two passes the lead vehicle and establishes the position for the center of the line. At the same time, the 3d and 4th vehicles swing out to the left and right. They form to the left and right and slightly to the rear of vehicle number two. The 1st and 4th squads of the lead platoon then move forward and form lines to the left and right, on the front end of vehicle number two. As soon as the troops are in position, the 3d and 4th vehicles close in on the squads and tighten the formation. The 2d and 3d squads of the lead platoon then form a line to the left and right of the 3d and 4th vehicles to complete the company line. The command vehicle (vehicle number one) takes

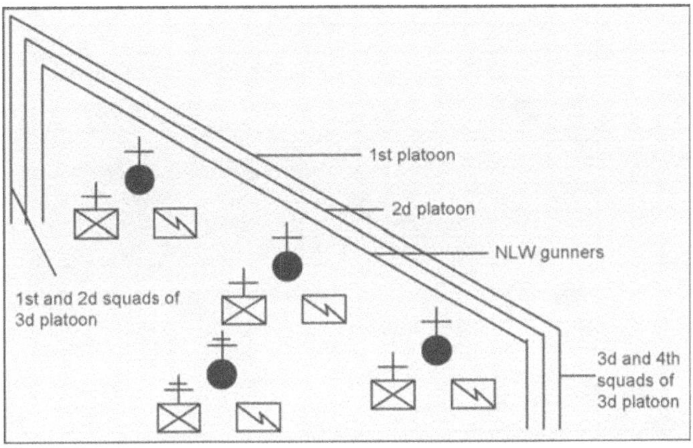

Figure 6-43. Company Echelon Right in Mass Formation With One Platoon in Lateral Support and One Platoon in General Support

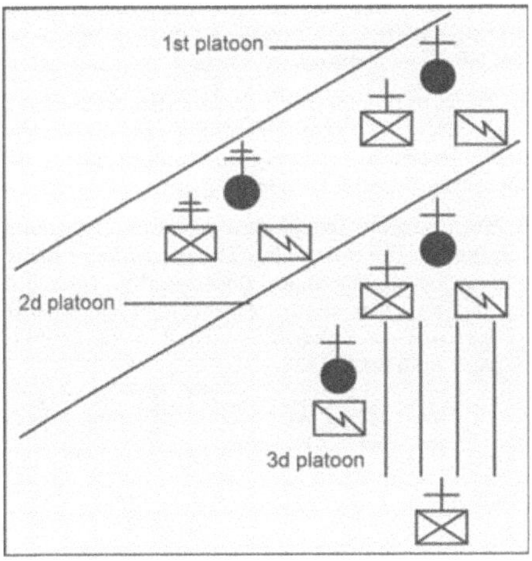

Figure 6-44. Company Echelon Right in Mass Formation With One Platoon in Lateral Support

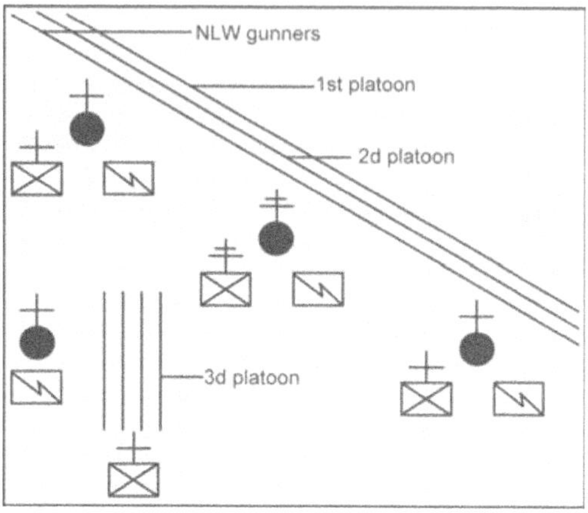

Figure 6-45. Company Echelon Right in Mass Formation With One Platoon in General Support

up a position behind the line where the commander can direct and control the unit. The 2d and 3d platoons move forward and are used in general, lateral, or direct support. In some instances, foot troops move into position first and vehicles join the formation, as needed.

6-105. To form a company echelon or a company wedge with vehicles, the company follows the same procedures used for company formations without vehicles. Vehicles are moved into the appropriate positions using procedures similar to those used for a company line.

6-106. To assemble troops and vehicles, the command vehicle moves to a position behind the formation. The commander faces the formation and gives the vehicle section the signal for assembly. Immediately, the vehicles return to their proper positions in column formation behind the command vehicle while the troops stand fast. The second in command then assembles the troops in the usual manner.

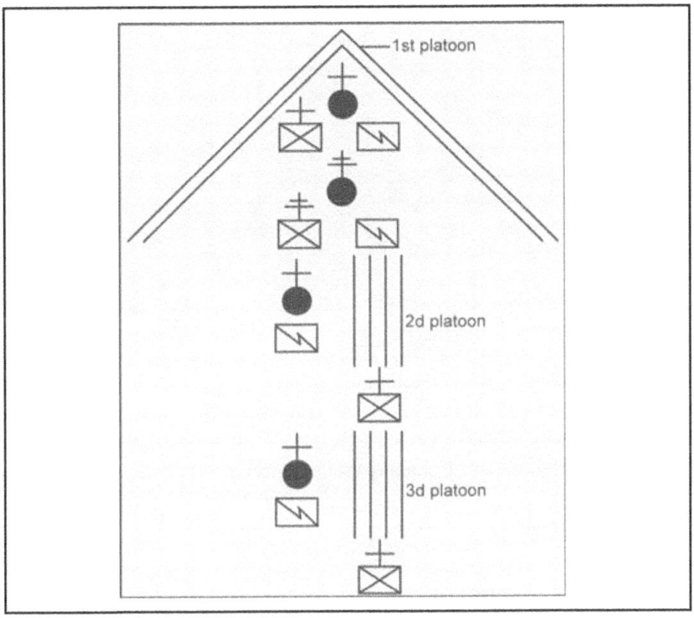

Figure 6-46. Company Wedge Formation With Two Platoons in General Support

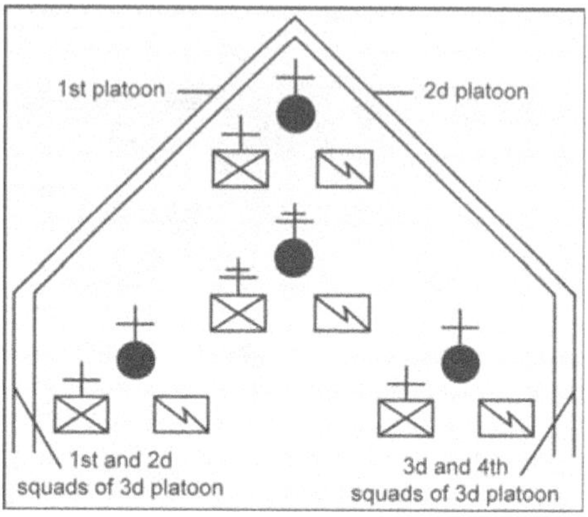

Figure 6-47. Company Wedge Formation With One Platoon in Lateral Support

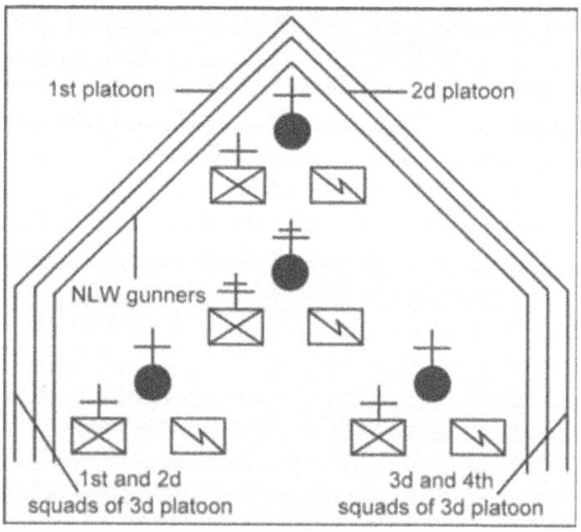

Figure 6-48. Company Wedge Formation With One Platoon in Lateral Support and One Platoon in General Support

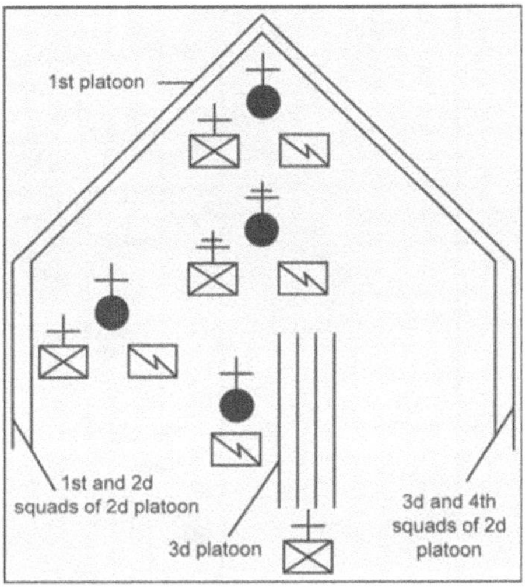

Figure 6-49. Company Wedge Formation With One Platoon in Lateral Support and One Platoon in General Support

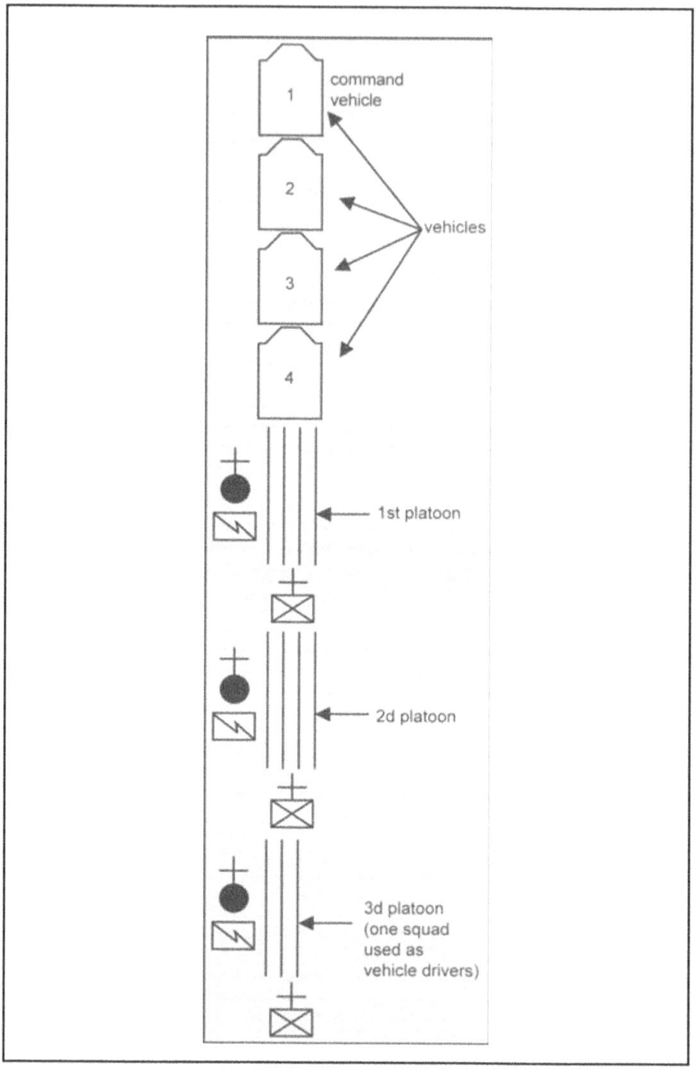

Figure 6-50. Company Column Formation With Vehicles

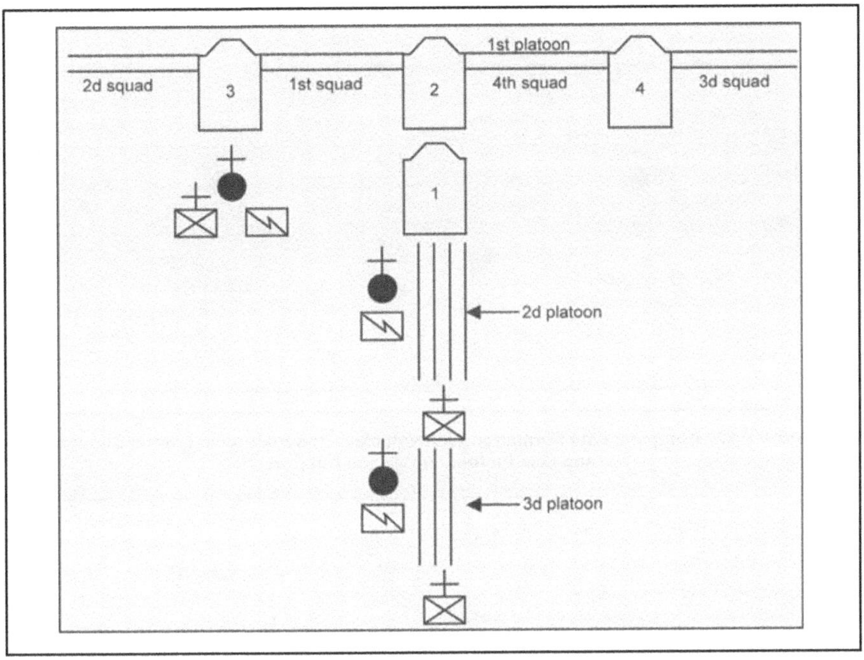

Figure 6-51. Company Line Formation With Vehicles and Two Platoons in General Support

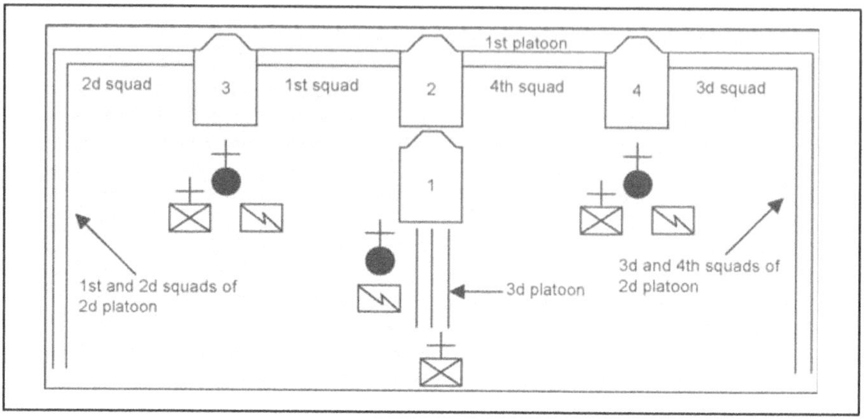

Figure 6-52. Company Line Formation With Vehicles, One Platoon in Lateral Support, and One Platoon in General Support

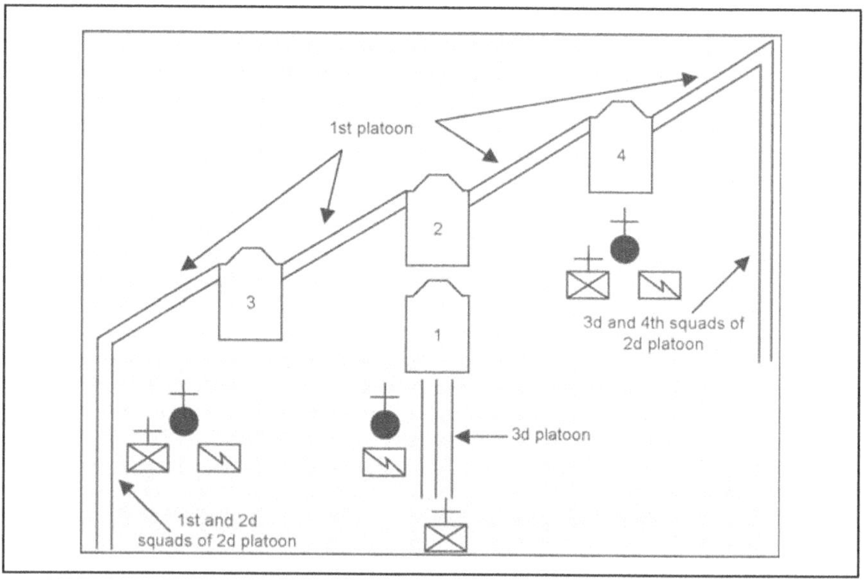

Figure 6-53. Company Echelon Right Formation With Vehicles, One Platoon in Lateral Support, and One Platoon in General Support

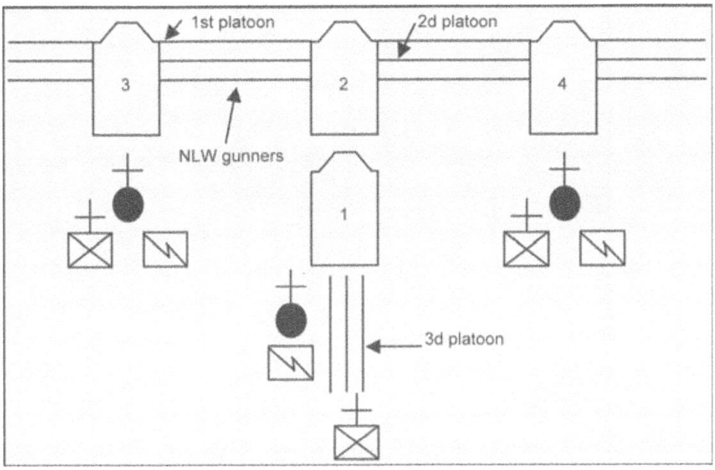

Figure 6-54. Company Line in Mass Formation With Vehicles and One Platoon in General Support

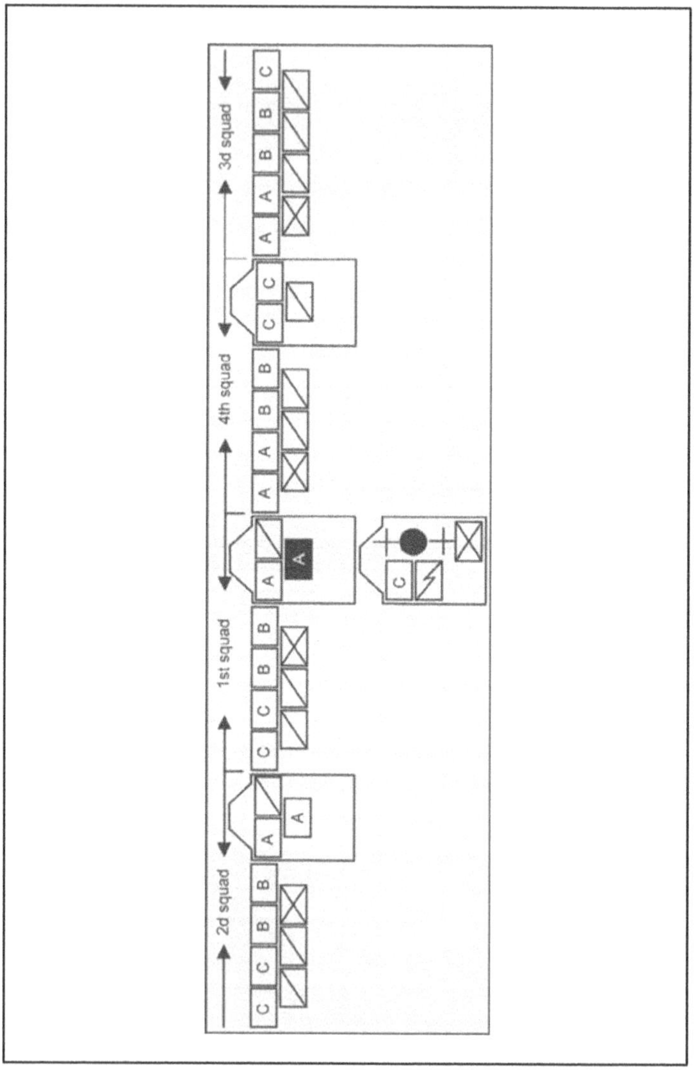

Figure 6-55. Platoon Line Formation With Vehicles

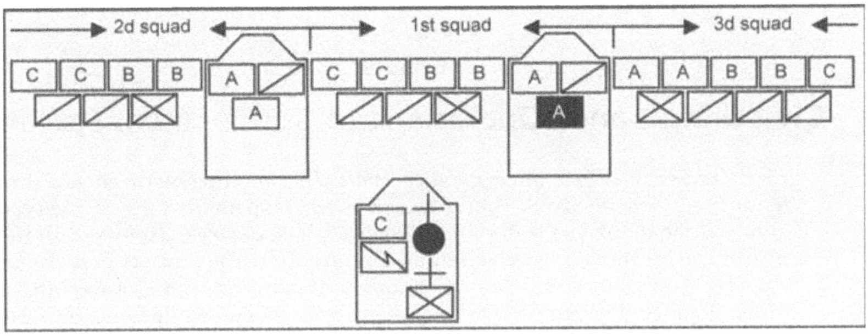

Figure 6-56. Platoon Line Formation With Direct Support and Vehicles

Chapter 7

Civil Disturbance Operations in Confinement Facilities

Civil disturbance doctrine is usually written for quelling disturbances that occur in large open areas where the size of the responding force or the type of formation employed is not hampered by a lack of space. However, in the military confinement environment, inmate disturbances often occur in small, closed areas, such as an inmate's housing unit, a gymnasium, a dining facility, or a chapel where maneuver space and equipment limit the size of the element that may be employed. The layout of each facility is different and presents its own unique problems when employing forces. This chapter covers the use of small military police teams, forced cell move teams (FCMTs), and the formations used to move an unruly inmate from one cell to another. Small generic riot control formations are discussed to provide the commander with a basic employment option that he can tailor to his facility to enable him to regain control of the affected area. The use of NL munitions and RCAs are incorporated into this discussion.

CROWD DYNAMICS WITHIN THE CONFINEMENT FACILITY

7-1. In confinement facility disturbances, inmates employ a number of tactics to resist control or to achieve their goals. Tactics may be unplanned or planned and nonviolent or violent. When a disturbance carries many purposes, it is likely that their tactics are well planned.

NONVIOLENT TACTICS

7-2. Nonviolent tactics range from name-calling to building barricades. Inmates may attempt to distract control force members by shouting at or ridiculing them or by using abusive language, obscene remarks, taunts, and jeers. The inmates' goals are to anger and demoralize the control force. They also want authorities to take actions that may later be exploited as acts of brutality.

VIOLENT TACTICS

7-3. Violent crowd tactics used by inmates are often extremely destructive and can include physical attacks on cadre, other inmates, and property. Use of violent tactics is limited only by the attitudes and ingenuity of the inmates, the training of their leaders, and the materials available to them. Inmates often commit violence with crude homemade weapons. If unplanned violence occurs, inmates will use mops, brooms, chairs, beds, or whatever else is on hand as weapons of violence. During planned violent disturbances, inmates can easily conceal makeshift weapons or tools to use against the control force. Inmates often erect barricades to impede troop movement and/or to prevent the control force from entering certain areas or buildings. Inmates can be expected to vent their emotions on individuals (cadre or inmates), the control force, their equipment, and/or the institution's property. Inmates direct dangerous objects like carts, barrels, liquids, and burning shoe polish at the control force.

GRIEVANCE PROTEST

7-4. A grievance protest may be organized as a riot. However, under normal circumstances, a riot of this type is not of an extremely violent nature. It may turn violent when leaders attempt to exploit any successes or weaknesses found in the control force.

UNORGANIZED RIOTS

7-5. Unorganized riots are characterized as being spontaneous in nature. However, they could be exploited and diverted by leaders at any subsequent stage of the riot, turning it into a different type. They may begin as an isolated assault against authority figures, or acts of civil disobedience. Under determined leadership, an unorganized riot could change to an organized riot.

MULTIPLE RIOTS

7-6. Multiple riots may happen when a disturbance occurs in one area of the confinement facility, causing others to riot in their areas. To control this situation with only one riot control force available, the force commander should first subdue the most violent riot. At the same time, he should employ some of his control forces to contain the other riots until the main control force is able to move to those locations to subdue them.

CADRE BEHAVIOR

7-7. The cadre is susceptible to crowd behavior. They become emotionally stimulated during tense confrontation with unruly and violent inmates. Commanders must be aware of the mood and attitude of the crowd and its effect on the control force cadre. To counteract the effects of crowd behavior on the cadre, the commanders must institute rigorous training and firm and effective leadership. This training must include a complete awareness and understanding of the use of force and the commander's intent. These are all necessary to offset the effect of crowd contagion upon the control force. Even with the best training and leadership, control force members must exercise individual and collective restraint.

PLANNING

7-8. Preplanning is the preparation and training conducted before a crisis occurs. The purpose of preplanning is to have plans and standing operating procedures (SOPs) in place so the cadre may react to an emergency and successfully contain and neutralize the situation. The preplanning process includes plan development, intelligence gathering, and training.

NOTE: Intelligence gathering is essential for the successful containment, neutralization, and/or prevention of a riot.

7-9. Once the control force team has been alerted to a disturbance within the confinement facility, leaders further develop preplans to fit the situation much the same way that DOD develops war plans.

7-10. During the planning process, leaders must be aware of the potential risks involved in quelling disturbances within a confinement facility. Careful planning minimizes collateral damage and risks to control force members and hostages.

7-11. The use of NLW and NL munitions must also be considered when developing plans. There must be strict accountability and control of RCAs, their employment, and other appropriate NL means (such as high-pressure water hoses). When using RCAs, plans must be flexible enough to accommodate changes in the situation and weather because it impacts the effectiveness of some agents. Planners must also consider the—

- Location of the disturbance.
- Estimated number of rioters.
- Accessibility of weapons, tools, or cleaning supplies.
- Inmate's military training.
- Hostage situation.

7-12. Based on an analysis of these factors, the commander makes an estimate of the situation. The estimate must be as thorough as time permits. Using the estimate, the commander considers COAs, selects RCAs, and determines the need for engineer support. In choosing a COA, consider the—

- Number of hostages.
- Number of inmates holding the hostages.
- Leader.
- Attitude and demeanor of the rioters.
- Location of the riot.
  - Inside areas of the confinement facility, such as the dining facility, work site, or housing wing.
  - Outside areas of the confinement facility, such as in exercise or work areas.
- Accessibility of weapons or material to make weapons.
- Accessibility of chemicals, such as cleaning supplies, cooking oils, and shoe polish.

USE OF CHEMICAL IRRITANTS

7-13. The use of chemical irritants can be a valuable NL tool for control force leadership to consider during the planning phase. These chemical irritants can drive a threat from an established, enclosed position or deny the rioters access to a certain area without long-lasting effects to those involved. The proper use of chemical irritants may prevent the control force from having to enter a dangerous area; however, improper use can cause injury, death, or property damage.

USE-OF-FORCE GUIDELINES

7-14. The installation commander authorizes the use of chemical irritants after the control force leader determines the type and dosage. The SOP outlines the procedures for securing authorization and provides guidance to help the control force leader make his decisions. The confinement facility commandant should inform the installation commander of the situation, since the installation commander is ultimately responsible. Team members must be aware of how chemical irritants affect personnel and plan accordingly. Inmates not involved must be removed from the area and the chemical cloud's path. The fire department and emergency medical services (EMS) should be on standby during operations involving chemicals.

PEPPER IRRITANT

7-15. The pepper irritant receives its name from varieties of the capsicum plant that is it made from. The active ingredient, Oleoresin Capsicum (OC), is extracted from the plant and micropulverized to make the irritant. It is neither a chemical nor a gas, but an all-natural organic substance. It is effective against individuals who are emotionally disturbed or under the influence of drugs or alcohol. It is effective against domestic and wild animals. OC is often used to divert the threat from making or continuing an assault. OC does not cause permanent damage. Its effects last 30 to 40 minutes after individuals are placed in fresh air.

7-16. OC can have various effects on people. Some effects are—

- Swelling of mucous membranes and the upper respiratory system; however, it does not shut down the system.
- Burning (intense) and discoloration (bright red) of exposed skin.
- Dilating the capillaries.
- Swelling of the eyelids.
- Burning and involuntary closing of the eyes.
- Coughing uncontrollably.
- Gagging.
- Gasping for air.
- Losing strength and coordination (temporarily).

OC has some disadvantages. They are that—

- The canister depressurizes over time if not checked regularly.
- Inmates may gain access to OC and use it against SRT members.
- The canister must be shaken on a regular basis.

DISSEMINATION OF CHEMICAL IRRITANTS

7-17. There are six methods for disseminating chemical irritants.

Pyrotechnic Method

7-18. In the pyrotechnic method of dissemination, the chemical irritant is placed in a canister with an inert material that is ignited when the device is activated. The agent is then carried into the air on the smoke particles of the inert material.

7-19. There are some disadvantages when using a pyrotechnic device indoors. These disadvantages are that the device—

- May cause a fire.
- Has a very slow saturation time (40 to 45 seconds to burn completely).
- Can be thrown back by the threat, even while burning.

Bursting Method

7-20. In the bursting method of dissemination, the irritant and an inert powder are placed in a saw-toothed canister. After a small detonation, the canister splits and expels the irritant in a cloud.

7-21. The bursting method has its advantages and disadvantages. Some advantages are that—

- There is no risk of fire.
- There is no risk of the canister being thrown back.
- The contents disseminate quickly.
- The canister is effective, easy to carry, and easy to control.

Some disadvantages are that—

- There is a slight possibility of fragmentation when detonated.
- The fuse head may separate from the canister.
- Only 50 to 90 percent of the irritant may be disseminated.
- The direction of the cloud formation depends on where the device is detonated.

Aerosol Method

7-22. In the aerosol method of dissemination, the irritant is suspended in an inert liquid that is located in the rear portion of the device. When the projectile penetrates the structure, the rear portion splits open and the irritant is disseminated in a mist. The aerosol method is most appropriate for tactical operations when chemical irritants are used before entry.

7-23. The aerosol method has its advantages and disadvantages. Some advantages are that it—

- Is easily carried.
- Is an excellent deployment system (uses 40-millimeter, 37-millimeter, or 12-gauge ammunition).
- Disperses the contents quickly.
- Is easily controlled and accurate (fin-stabilized).

NOTE: Practice firing is always necessary.

Some disadvantages are the possibility of—

- Injury to individuals if it is fired from close range.
- Damage to the structure if it strikes a weak portion.

Projectiles Method

7-24. Chemical-containing projectiles can be very effective when used correctly. A soldier must know which projectile to use in each unique situation.

7-25. **Thirty-Seven-Millimeter Projectile.** A 37-millimeter projectile can incapacitate anyone in an enclosed area of 4,500 cubic feet. It can penetrate the following:

- A sheet of ¾-inch-thick plywood at 25 yards.
- An automobile windshield at 33 yards.
- A hollow-core door at 58 yards.

7-26. **Forty-Millimeter Projectile.** A 40-millimeter projectile is more effective at greater ranges than a 37-millimeter projectile because of its rifled bore. It can incapacitate anyone in an enclosed area of 4,500 cubic feet. It can penetrate the following:

- A 1-inch-thick plywood at 10 yards.
- An automobile windshield at 55 yards.
- A hollow-core door at 109 yards.

7-27. **Twelve-Gauge Projectile.** A 12-gauge projectile can be loaded into a shotgun with no modifications to the weapon. It can be fired as a single round, or it can be magazine-fed. The 12-gauge projectile can incapacitate anyone in an enclosed area of 1,000 cubic feet. The angle at which the projectile strikes the barrier can affect its penetration capability. It can penetrate the following:

- A sheet of ¾-inch plywood at 33 yards.
- An automobile windshield at 33 yards.
- A hollow-core door at 100 yards.

Fogging Method

7-28. The irritant is suspended in a fogging solution and dispersed with a fogging machine. It is not recommended for tactical operations.

7-29. The fogging method has its advantages and disadvantages. Some advantages are that it—

- Produces enough chemical irritant to cover an enclosed area of 100,000 cubic feet in 26 seconds.
- Disseminates quickly.
- Provides high-volume capabilities.
- Is effective for riot control or other crowd dispersal situations.

Some disadvantages are that it is—

- Noisy.
- Cumbersome.

Multipurpose Grenade Method

7-30. A multipurpose grenade can be hand-thrown or launched from a shotgun with an adapter. It has an extended shelf life of 6 years and an adjustable fuse delay of 2 to 5 seconds. The irritant is located in the cylinder portion of the device and is forced out the bottom of the grenade upon detonation.

WARNING
A multipurpose grenade deployed incorrectly could result in injury. Safety rules should be observed at all times.

RECORD OF EVENTS

7-31. Reporting procedures for serious incidents should be included in the plans. A record of events must be initiated to provide a basis for the preparation and submission of a formal report to higher HQ. As a minimum, include the—

- Time the incident was reported and by whom.
- Time the incident was reported to the facility commander.
- Time the control force was assembled.
- Time the control force entered the facility.
- Weather conditions as they relate to the use of RCAs.
- Number of cadre and inmates injured or killed, including how they were injured or killed and the medical attention given to them.
- Time the operation was completed.
- Time the riot control force restored order.

TRAINING

7-32. Guard units and associated teams must train on a regular basis in the five basic riot control formations. There must be a continuous training program established, to include—

- Interpersonal communication.
- Self-defense.
- Use of force.
- Use of a riot baton.
- Fatal areas of the body to avoid when using the riot baton.
- First aid.
- Emergency plans.
- Intelligence and counterintelligence techniques.
- The use of RCAs and the various methods of dispersing them.
- The use of NLW.
- The use and effects of high-pressure water.
- Riot control formations.

EQUIPMENT

7-33. The recommended basic riot gear used by the confinement riot control force is as follows:

- Riot baton.
- Riot shield.
- Helmet with face shield.
- Groin protector.
- Flak vest.
- Leather gloves.
- Shin protection.

RIOT BATON

7-34. The riot baton is an invaluable weapon in crowd control situations. A riot baton in the hands of a well-trained control force member is the most appropriate weapon, except with extremely violent crowds. The riot baton is not meant to replace NL munitions, but is instead used in situations in which NL munitions are not needed. Losing a riot baton to the crowd does not present a serious threat.

7-35. Use of the riot baton is based on the commander's appraisal of the situation and his choice of force option. Control forces committed with the riot baton must have teams equipped with RCAs, NLW, and lethal coverage. If the confrontation increases in intensity, the commander has these force options available and may rapidly engage the control force.

7-36. A riot baton is a versatile offensive weapon that varies in length from 26 to 42 inches with the optimum length being 36 inches. It is made of hickory and does not shatter or break easily. Both ends of the riot baton are rounded to prevent unnecessary injury. It is approximately 1¼ inches in diameter. Each riot baton has a hole drilled at the grip end with a leather thong threaded through the hole. The thong helps secure the riot baton to the user's hand.

RIOT SHIELD

7-37. The riot shield (see *Figure 7-1*) is an excellent piece of defensive equipment. It is used by control force personnel in the front and on the flanks of the formation.

7-38. The riot shield has its advantages and disadvantages. Some advantages are that—

- OC pepper spray or high-pressure water can be used to back the control force.
- It provides protection from thrown objects.
- It absorbs some of the impact of rioters hurling themselves at the control force.

The disadvantages of the riot shield are that it—

- Severely limits the use of the riot baton.
- May crack if hit with a heavy object.

Figure 7-1. Riot Shields

RIOT CONTROL FORMATIONS

7-39. The five basic formations for civil disturbance operations are the line, wedge, echelon, diamond, and circle. In the correctional environment the basic formations for disturbances are the line, wedge, or echelon with apprehension teams in support. Generally, the diamond and circle formations are not used in the correctional environment. The minimum size force that the commander should consider is an augmented platoon, with a two-to-one advantage over the rioters.

7-40. Correctional facility cadre must be proficient in all riot control formations. These are usually trained at the squad level and above, but performed by a platoon or company-sized unit. Squad line, echelon left, echelon right, and wedge formations form the basis for platoon and company formations. Each squad must be adept in the basic formations before practicing in platoon-sized or larger formations. For more detailed information on formations, see *STPs 19-95B1-SM* and *19-95B24-SM-TG*. In the correctional environment, the support platoon plays a key role by supporting the control formation with apprehension and equipment teams. See *Chapter 6* for an in-depth discussion on civil disturbance formations.

APPREHENSION TEAMS

7-41. Apprehension teams give the control force the ability to extract inmates. Inmates identified as leaders or agitators and inmates that are injured or no longer want to resist can be extracted from the crowd. Apprehension teams can also be used to enter confined areas where the formation cannot operate.

7-42. The apprehension team will consist of eight members and one leader. Each team will have four shield holders, two nonlethal gunners and two apprehension/search members. The squad leader controls the movement of the team and assists the team in exiting the control formation. He maintains communications with the formation and coordinates the team's operations with the formation. The squad leader also ensures that the team uses the minimum amount of force necessary to extract inmates.

7-43. The apprehension team may deploy through the middle of the formation or from around either of the flanks.

7-44. The apprehension team should be deployed within a distance of approximately 10 meters so that the formation can still provide immediate support. If the apprehension team is deployed farther than 10 meters, the formation must advance to close the gap.

7-45. The team will exit the formation in two columns; shield holders in the first and second ranks, nonlethal gunners in the third rank, apprehension/search members in the fourth rank and the leader last. Team members grip the shoulder or equipment of the member in front of them. The shield holders will envelop the inmate that is to be extracted and form a protective barrier. Once the shield holders have enveloped the inmate, the nonlethal gunners will cover their sides with their weapons while the apprehension/search members control and restrain the inmate. When the team has the inmate under control, the squad leader will give the order for the team to withdraw. The team withdraws in reverse order (apprehension/search members with the inmate, followed by nonlethal gunners and shield holders). Team members then move backwards, face the crowd and grip the shoulder or equipment of the member in front of them.

TEAM EQUIPMENT

7-46. The apprehension team uses the following equipment:

- Flex-Cufs.
- Restraint cutters.
- Pepper spray.
- Helmets with face shields.
- Groin protectors.
- Flak vests.
- Elbow pads.
- Leather gloves.
- Shin protectors.
- Knee pads.
- A video camera.

TEAM COMPOSITION

7-47. The apprehension team is composed of five team members. This team consists of four members and a squad leader. Team member one is the team leader.

Squad Leader

7-48. The squad leader assists the team in exiting the control formation and ensures that the team uses the minimum force necessary to subdue inmates. He is also responsible for signaling the PSG if more help is needed.

Team Member One (Team Leader)

7-49. Team member one's responsibility is to make initial contact with the inmate and use pepper spray and apply Flex-Cufs, as needed. He also carries a set of restraint cutters. Once the apprehension team has returned to the relative safety of the control formation, team member one searches the inmate.

7-50. Team member one receives the command to move forward and secure the inmate from the squad leader. This command is usually given by a signal and verbal command. Team member one then notifies the main force commander where his team exits the control force. Once the team is in position to exit the main formation, team member one signals the member of the formation in front of him. This is usually a prearranged signal, such as a tap on the right shoulder. The control force opens and the team exits the formation and makes contact with the assigned inmate. Team members then subdue and control the inmate. If pepper spray is needed, team member five shouts, "Pepper." This gives the other team members time to react to the command and spray the inmate. Team member one then applies Flex-Cufs to the inmate's wrists and ankles. Depending on the mission, the inmate may be left for follow-on security forces or carried behind the control force to be secured by security forces.

7-51. Should the mission require that the inmate be moved, team member one shouts, "Prepare to Move," this allows each member to grab an arm and position the inmate toward the control force formation. When the team is ready to move, team member two shouts, "Ready," alerting team member one that the team is ready for the last command. Team member one then commands, "Move." The team enters the control formation with the inmate. The inmate is searched by team member one and turned over to the security team.

Team Member Two

7-52. Team member two stands directly behind team member five with his hands on team member five's waist. He protects the left side of team member five by moving to team member five's left and helps to secure the inmate after team member one makes initial contact. He repeats team member five's commands and controls the inmate's head after the inmate is on the ground.

Team Member Three

7-53. Team member three stands behind team member two with his hands on team member two's waist. He protects the right side of team member five by moving to team member five's right and helps to secure the inmate after initial contact.

Team Member Four

7-54. Team member four stands behind team member three with his hands on team member three's waist. His task is to help subdue the inmate in any way possible. He usually concentrates on the inmate's body below the knees. He acts as the eyes and ears for the team during control of the inmate.

Team Member Five

7-55. Team member five gives commands to control the inmate.

FORCED CELL MOVE TEAMS

7-56. The intent of a forced cell move is to move an unruly and/or uncooperative inmate from one cell to another. This is a difficult task and must not be taken lightly. A forced cell move must be thoroughly planned, rehearsed, and resourced to ensure the safety of the team members and the inmate.

7-57. A number of reasons may compel the commander to consider a forced cell move. The inmate may refuse to eat, take medication, accept medical attention, get a haircut, keep himself clean, work, or move or rotate cells when ordered to. These are some of the reasons that the commander may order a forced cell move.

TEAM COMPOSITION AND EQUIPMENT

7-58. The FCMT is composed of six corrections cadre and two support cadre from military police and medical sections specially trained in forced cell movement. Primarily, the team consists of the following personnel:

- One officer in charge (OIC) (E7 or above).
- Five corrections cadre.
- One military police investigator.
- One medic.

7-59. The FCMT uses the following standard equipment:

- Helmets with face shields.
- Groin protectors.
- Flak vests.
- Elbow pads.
- Leather gloves.
- Shin protectors.
- Knee pads.
- A pinning shield.
- Restraints (hand or leg irons or Flex-Cufs).
- A video camera.

ACTIONS PRIOR TO TEAM EMPLOYMENT

7-60. The OIC gives an operations order (OPORD) to team member five (the team leader) of the FCMT. As a minimum, the OPORD should answer the following questions about the incident leading up to the move:

- Did the inmate fail to comply with orders?
- Did the inmate assault cadre or other inmates?
- Is the objective of the move to force a shower or haircut?
- Is OC pepper spray authorized for use?
- Has the inmate been moved before?

- What is the condition of the area where the inmate is currently located?
- Has the inmate tested positive for human immunodeficiency virus (HIV)?
- Does the inmate have a weapon or the materials needed to make one?
- Is the area barricaded?
- Has he smeared himself with body fluids or other slippery substances?
- Can you approach the inmate without risk to him and/or others?

7-61. The team is assembled for inspection by the OIC, and the team members are required to brief the OIC on their individual duties and responsibilities. A line formation is the customary formation for this inspection. It should take place out of site of the inmates.

7-62. Each soldier takes one step forward and briefs the OIC as shown in the following example: *"I am (rank and name), team member one. My mission is to pin the inmate using the minimum amount of force necessary. I carry the pinning shield."* He then faces so the camera can record the number printed on the back of his flak jacket and holds up the pinning shield [see *Figure 7-2*]).

Figure 7-2. Recording Team Members' Duty Position

INDIVIDUAL TEAM MEMBER RESPONSIBILITIES

7-63. Each individual team member has his own responsibilities during a forced cell move. Each team member must do his designated job while working as a team.

Team Member One

7-64. Team member one is the eyes and ears of the team. His responsibility is to carry the pinning shield up to the cell door and describe the scene with as much detail as possible (especially including anything that may cause harm to the team). The pinning shield is used to protect the team from body fluids thrown at the them by the inmate.

7-65. Team member one describes the scene in the cell to his teammates by turning his head to the right and speaking in a loud voice so that all team members can hear him. A slap on the right shoulder by team member two alerts him that all team members have heard him. If he receives a slap on the left shoulder from team member two, he repeats the situation report (SITREP) in a loud, clear voice. The decision to enter the inmate's area with or without the pinning shield is team member one's decision, unless the inmate is known to have a weapon. Should he decide to enter without the pinning shield, team member one then enters with both hands held approximately chest high. When team member one is unable to determine if the inmate has a weapon, the pinning shield is used. Charging through the open cell door, team member one drives the inmate to the bunk or ground with the shield. Once pinned, team member one secures the inmate's head and ensures that it is not twisted or turned in relation to its natural form. He may have to use the various pressure points on the inmate's head to help subdue him.

Team Member Two

7-66. Team member two steps forward and briefs the OIC as follows: *"I am (rank and name), team member two. My mission is to secure the inmate's right arm using the minimum amount of force necessary. I carry one set of hand irons."* He then holds the hand irons up for video camera taping (see *Figure 7-3*) and says, *"I also carry one canister of OC pepper spray to be used at the direction of the team OIC."* He then holds up the canister for video camera taping and faces so the camera can record the number on the back of his flak jacket (see *Figure 7-2,* page 7-13.)

Figure 7-3. Recording Team Members' Equipment

7-67. Team member two lines up directly behind team member one. He keeps contact with team member one until he can secure the inmate's strong arm. Team member two taps team member one on his right or left shoulder after being tapped by team member three. Once inside the cell, team member two secures the prisoners strong hand with wrist restraints and announces, "*Hands secure,*" to team member five.

Team Member Three

7-68. Team member three steps forward and briefs the OIC as follows: *"I am (rank and name), team member three. My mission is to secure the inmate's left arm using the minimum amount of force necessary."* He then faces so the camera can record the number on the back of his flak jacket (see *Figure 7-4*).

Team member three lines up directly behind team member two. Team member three keeps contact with team member two until he can grasp a portion of the inmate's body. He then concentrates on securing the inmate's weak hand and assisting team member two in applying hand restraints. Team member three taps team member two on his right or left shoulder after being tapped by team member four.

Figure 7-4. Recording Number on Flak Jacket

Team Member Four

7-69. Team member four steps forward and briefs the OIC as follows: "*I am (rank and name), team member four. My mission is to secure the inmate's right leg using the minimum amount of force necessary. I carry one set of leg irons.*" He holds the leg irons up for the camera to record (see *Figure 7-2,* page 7-13). He then faces so the camera can record the number on the back of his flak jacket (see *Figure 7-2*). Team member four lines up directly behind team member three. He taps team member three on his right or left shoulder after being tapped by team member five. Team member four keeps contact with team member three until he can secure a portion of the inmate's body. Team member four concentrates on securing the right leg of the inmate. It is his responsibility to apply the leg irons. Once the legs are secure, he announces, "*Legs secure,*" to team member five.

Team Member Five (Team Leader)

7-70. Team member five steps forward and briefs the OIC as follows: *"I am (rank and name), team member five. My mission is to secure the inmate's left leg using the minimum amount of force necessary. I carry a restraint key (hold up for the camera to view). I am also the team leader for this mission and it is my responsibility to ensure that the team uses the minimum amount of force necessary to subdue the inmate."* He then faces so the camera can record the number on the back of his flak jacket (see *Figure 7-4*).

7-71. Team member five taps team member four on the right shoulder when he has heard the SITREP from team member one. If he does not hear the report or needs to hear it again, he taps team member four on the left shoulder. Team member five keeps in contact with team member four until he can secure a portion of the inmate's body. He secures the left leg and assists team member four in applying the leg irons. Once he has heard that the hands and legs are secure, team member five signals for the team to conduct an equipment check. The team members check the equipment in place.

NOTE: At no time is the inmate to be released by anyone.

Medic

7-72. The medic briefs the OIC in the following manner: *"I am (rank and name), the medic. My mission is to treat all injuries that may occur during the move."* He follows the instructions given to him by the OIC/NCOIC. His mission is to observe the inmate for signs of injury before the forced cell move and assess the inmate's condition during and after the move.

Military Police Investigator

7-73. The military police investigator (MPI) is responsible for videotaping the entire forced cell move. This includes the OIC/NCOIC briefing the OPORD, each team member briefing his responsibilities to the OIC, and any medical treatment rendered to the inmate.

ACTIONS DURING THE FORCED CELL MOVE

7-74. There are many actions performed during a forced cell move. Each member of the FCMT must complete his task successfully to accomplish the forced cell move.

Officer in Charge or Noncommissioned Officer in Charge

7-75. The OIC, an E7 or above, is responsible for ensuring that the team uses the minimum amount of force necessary to subdue the inmate. He is also responsible for ensuring that all necessary precautions have been taken to ensure the safety of the team and the inmate. The OIC gives the inmate the initial orders to comply and informs him of the consequences of noncompliance. He does this by placing himself at the front of the inmate's cell (see *Figure 7-5*). The orders are not given until it can be enforced by the FCMT. The following prepared statement is read to the inmate by the OIC: *"I am (rank and name), OIC of this FCMT. At this time I am ordering you to lie face down on your bunk (or other designated location) with your head to the rear of the cell. Cross your hands behind your back and cross your feet as the team enters the cell and uses the minimum amount of force necessary to place you in restraints and move you to (give location). Failure on your part to comply with these instructions or to resist the FCMT will result in the use of force (OC pepper spray if applicable). Do you understand these instructions? If so, lie down on the bunk."*

7-76. After reading the instructions to the inmate, the OIC faces the camera and states whether the inmate complied. He then waits 60 seconds and motions for the team to position itself in front of the inmate's cell (see *Figure 7-6*). The OIC directs the opening of the cell when the team is positioned at the entrance to the cell and prepared to enter. He positions himself so he can observe team entry to and departure from the cell. If OC spray is authorized, the OIC tells the team when to spray the inmate with OC spray. This task is done in silence. Communication between team members should be done through hand-and-arm signals, unless noted otherwise.

Figure 7-5. OIC Reading Compliance Statement to Inmate

Figure 7-6. OIC and FCMT

Team Member One

7-77. Once the team is positioned in front of the inmate's cell, team member one starts his report. At a minimum, he must answer the following questions and report them to the team:

- Does the inmate have a weapon, if so, what kind?
- Does the inmate appear dry?
- Is the inmate clothed or wearing extra clothing?
- Has the inmate smeared anything on his body?
- Can you see the inmate's hands?
- Is the cell barricaded or visibly booby-trapped?

- Is the cell dry?
- Are cell furnishings intact?

7-78. Based on the answers to the above questions, team member one must decide whether to enter the cell with or without the shield (see *Figure 7-7*). Team member one's ultimate goal is to pin the inmate on the bunk or floor.

Figure 7-7. FCMT Preparing to Enter a Cell

Team Member Two

7-79. Team member two keeps contact with team member one until he can secure the inmate's strong arm. If he cannot immediately secure the inmate's arm, he secures a portion of the inmate's body and progresses to the inmate's strong arm. Once the inmate's hands are secure, he announces, "*Hands secure*," to team member five (see *Figure 7-8)*.

Team Member Three

7-80. Team member three keeps contact with team member two until he can grasp a portion of the inmate's body. Team member three concentrates on securing the inmate's weak hand and assists team member two in applying Flex-Cufs. Team member three maintains control of the inmate's hands and looks at team member five for further instructions (see *Figure 7-8*).

Team Member Four

7-81. Team member four keeps contact with team member three until he can secure a portion of the inmate's body. Team member four concentrates on securing the inmate's right leg. He applies leg irons and when the inmate's legs are secure, he announces, "*Legs secure*," (see *Figure 7-8*).

Team Member Five (Team Leader)

7-82. Team member five keeps contact with team member four until he secures a portion of the inmate's body. Team member five concentrates on securing the inmate's left leg and assists with the application of leg irons. When he is informed that the hands and legs are secure, the team searches the inmate for weapons and contraband. Once the search is complete, team member five commands, "Prepare to Lift."

Figure 7-8. FCMT Securing an Inmate

Movement of the Inmate

7-83. Team member one positions his hands to control the inmate's head during the move, turning the inmate's head toward the wall. Team members two and three position themselves on the inmate's right and left sides. They grasp the wrist or forearm of the team member opposite them to form a platform for the inmate to lie on. Team member four positions himself on the inmate's right side even with the inmate's knees. This places him in position to control the inmate's legs. Team member five's position is on the inmate's left side and enables him to control the inmate's lower legs and feet. When in position, team member five ensures that all team members are in place before commanding, "Ready, Lift."

7-84. On the command "Lift," all team members simultaneously stand with the inmate. They should sandwich the inmate between them to facilitate control. Team member five then commands, "Prepare to Turn, Ready—Turn." The inmate is always turned toward the bunk. If a bunk is not available, team member five should specify which direction to turn the inmate. The inmate is then removed from the cell (see *Figure 7-9*, page 7-20).

7-85. Once the inmate is on the tier, team member five commands, "Prepare to Lower, Team—Lower." The team simultaneously lowers the inmate to the floor with the team keeping positive control of the inmate. The medic asks the inmate, "*Do you have any injuries at this time?*" (see *Figure 7-10,* page 7-20). If the inmate has no injuries, the medic looks at the camera and states, "*Inmate (inmate name) appears to have sustained no injuries.*" If the inmate has sustained injuries, the medic treats those injuries and completes a sworn statement. If OC was employed, the medic decontaminates the inmate using water and paper towels. The medic instructs the inmate to tilt his head to the side and flush his eyes with water for approximately 5 minutes. The medic then dries the inmate's face.

Figure 7-9. Moving the Inmate

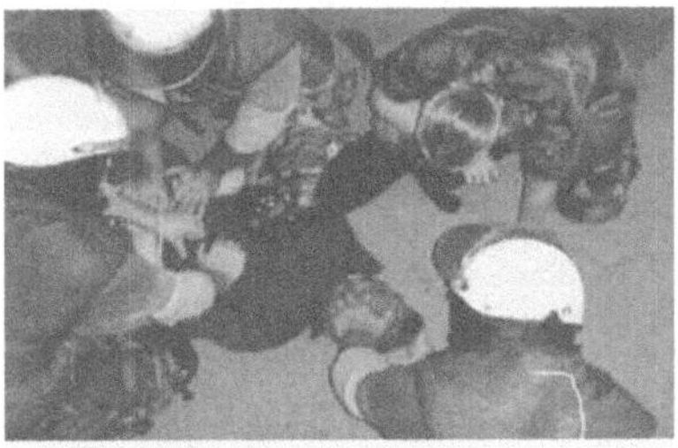

Figure 7-10. Treating Injuries

Placement of the Inmate in a Cell

7-86. Team member five orders an equipment check before returning the inmate to a cell. He then commands, "Prepare to Lift, Team—Lift." Team members position themselves exactly where they were when they removed the inmate from the cell. Next, team members simultaneously lift the inmate and return him to the predetermined cell. They move the inmate into the cell head first with his face toward the bunk. Once inside the cell, team member five commands the team to lower the inmate onto the bunk (see *Figure 7-11*, page 7-20). Team members then remove the inmate's restraints, keeping positive control of the inmate at all times. Team members use a nonverbal signal to notify team member five when the restraints are removed. He signals another equipment check. Team members then account for all equipment in place and prepare to exit the cell. The team exits the cell in the same manner as they entered, but in reverse order.

Figure 7-11. Placing Inmate in Cell

7-87. Team member four moves into position where he can control both of the inmate's legs while team member five moves toward the cell door. When team member four is ready to exit the cell, team member three moves to control the inmate's legs, team member two moves to where he can control both of the inmates arms, and team member one maintains control of the inmate's head. Team member four maintains contact with team member three at all times. When team member three is ready to exit the cell, team member two moves into a position where he can control the inmate's lower body and team member one moves into position to control the inmate's upper body. Using a prearranged, nonverbal signal, the team pulls team members one and two from the cell. The FCMT is prepared to reenter the cell if the inmate attempts to attack team members one and two as they exit the cell. The FCMT remains ready to reenter until the cell door is secured.

Chapter 8

Civil Disturbance Training

Training requires planning and preparation. Unit commanders structure their training program according to the mission essential task list (METL) (see *FM 7-1)* and expected contingencies. These training plans include prior training and the experience and proficiency of their unit. Leaders agree that the operational tempo and expected contingencies for the next 10 to 15 years will likely involve low-intensity military operations other than war (MOOTW). Therefore, civil disturbance training must be included in the commander's annual training program and must cover all aspects of that type of operation. To effectively function as a member of a control force, individual soldiers must be trained in civil disturbance operations. Training must be intensive, realistic, and sustained. Any lack of training (from basic soldier skills to leadership skills) can have an adverse effect to the point of endangering lives, especially the lives of soldiers. Leader involvement in the training process is essential for mission success in crowd control operations.

UNIT TRAINING STRATEGY DEVELOPMENT

8-1. Developing a unit training strategy begins with the mission analysis portion of the MDMP and identifying the GRM. Missions that require soldiers and units to enforce treaties or accords, protect the lives of civilians in uncertain or hostile environments (such as NEOs), or provide large-scale humanitarian assistance demand some sort of graduated-response criteria to maintain order and prevent uncertain environments from becoming hostile.

8-2. When developing training strategies, planners should focus training and evaluation on the echelon level above the target element (for example TF) or the unit identified as needing training.

8-3. Planners must agree on the intent of the GRM. The GRM is for most a training and rehearsal tool. It provides leaders with likely vignettes that can be incorporated into COA analysis, predeployment training, and rehearsals. The GRM becomes a handy reference during situations that require graduated responses.

TRAINING PLAN DEVELOPMENT

8-4. In developing a training plan for civil disturbances, many considerations should be addressed in terms of available resources. Considerations include the number of personnel to train, the equipment needed and its availability, the time available to train to standard, the location of where training will be executed, and the incorporation of SOPs (based on the theater of operations

[CONUS or OCONUS]). A personnel assessment is necessary to determine the training and experience level of soldiers. It is critical to identify and use soldiers with previous real-world experience and leaders that have graduated from the Interservice Nonlethal Individual Weapons Instructor Course (INIWIC) as primary instructors.

8-5. Commanders should structure the unit training program based on the METL, expected contingencies, prior training, and the proficiency of the unit. In developing a training program, use the seven-step approach. For in-depth instructions on training development, see *FMs 7-0* and *7-1*.

8-6. The objective is to train and evaluate units in the performance of civil disturbance procedures. Proficiency in performing civil disturbance procedures is reflected by a unit's ability to quickly and deliberately execute the disturbance control procedures required to prevent loss of life and property or to regain control of a civil disturbance in its assigned area of responsibility.

8-7. Training should focus on, but is not limited to—

- Instructors.
- Individual and collective drills.
- Leaders.
- Equipment.
- Weapons.
- Munitions.
- Live-fire range exercises.
- Qualification and familiarization.
- Soldier training exercises (STXs).
- Culminating exercises.

8-8. There are several successful methods to evaluate trained leader skills and trained soldier skills using scenarios, vignettes, lanes, or STXs. Flexibility is an important key in training plan development due to time restraints.

RANGE SETUP

8-9. The key to setting up a successful range for NL munitions is to ensure that they are in compliance with installation safety regulations. Kinetic energy is the force that pushes the projectiles down range. In most cases, kinetic energy will only carry these NL munitions projectiles 75 meters or less. This gives trainers flexibility as to the locations where a live-fire exercise can be conducted, especially in a field-expedient situation. *Figure 8-1* is an example for conducting a NL live fire range setup.

8-10. As illustrated in *Figure 8-1*, mark off 5-, 10-, 20-, 30-, 40-, and 50-meter distances from the targets with markers. Each of the NL munitions has minimum and maximum effective ranges within the distances of 5 to 50 meters. Impact power and accuracy are greater at close distances and depleted at farther distances. Targets can be anything from the standard E-type to more realistic three-dimensional manikins. Evenly divide your shooters into three groups as shown in *Figure 8-1*. Group one will initially be shield holders, group two will be shooters, and group three will be grenadiers.

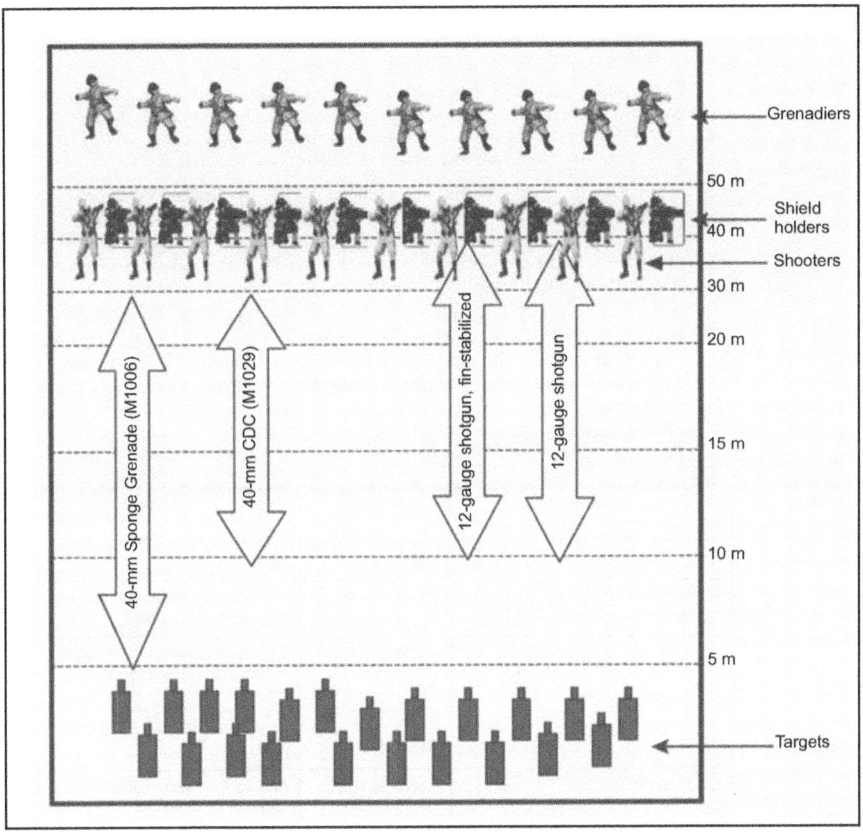

Figure 8-1. NL Munitions Range Setup

8-11. The formation should begin at the farthest effective range firing line (50 meters). After each volley of fire, the formation moves closer to the targets, but stays outside the closest effective range firing line (5 meters). For safety reasons, ensure that the formation stays online throughout each iteration.

8-12. When moving forward to each firing point, it is recommended that the formation always come to a halt before the fire command is given. It is also important to use the tap down drill method (see *Figure 8-1*) rather than having the shooters move forward and fire between the shield holders.

8-13. Range instructors can use verbal commands, such as, "Ready," or "Fire," for each volley. Shooters can be briefed to move forward, tap down the shield holder, and engage targets at the first explosion of distraction grenades thrown by grenadiers. In that instance, the command could simply be "Grenadiers Prepare to Throw, Throw."

8-14. The success of any civil disturbance operation is through effective communication, especially when incorporating the use of NL munitions. For effective training, use verbal commands in conjunction with effective, predetermined hand-and-arm signals during live-fire exercises.

8-15. Each iteration requires two phases of fire. During the first phase, half of the shooters will be firing NL, 40-millimeter rounds using an M203. The others will be firing NL, 12-gauge rounds using a 12-gauge shotgun. In the second phase, shooters switch weapons (safely). It is critical that line safeties are alert during this transition between phases.

8-16. Upon completion of each iteration, clear the weapons and the range. Establish a safe method and return all groups back to the farthest firing line. Once that is accomplished, group one rotates to become group three, group three rotates to become group two, and group two rotates to become group one for the next iteration. Using this method affords a safe and efficient way to get all shooters through the proficiency course. *Figure 8-2* is an example of how to fire each volley based on its characteristics, effectiveness, and capabilities.

Distance	Command	Subject	Actions
10 Meters	12-Gauge Shotgun, Two Point	Shotgun shooters	Load two M1013 rounds.
	M203, Two Point	M203 shooters	Load one M1006 round and have one readily available.
	Ready, Fire	Shotgun shooters	Tap down shields and engage designated targets.
		M203 shooters	Tap down shields, engage designated targets, reload, and reengage.
	Clear Range	Range safeties	Ensure that weapons are clear and safe.
	Safeties, Clear All Weapons	All shooters	Face to right, keep weapons pointed up and downrange.
	Shooters, Right Face, Move	All personnel	Step off shield holder, then shooters, and grenadiers on the command, "Move." Grenadiers reform at 50 meters.

Figure 8-2. Sample Live-Fire Exercise

Distance	Command	Subject	Actions
15 Meters	Shotgun, Two Point	Shotgun shooters	Load two M1013 rounds.
	M203, Two Point	M203 shooters	Load one M1006 round and have another available.
	Grenadiers, Prepare to Throw	Grenadiers	Make distraction grenade ready.
	Throw	Grenadiers	Throw grenade.
		Grenadiers	Throw distraction grenade.
	Ready, Fire	Shot gun shooters	Tap down shields and engage designated targets.
		M203 shooters	Tap down shields, fire at designated targets, reload with M1006 rounds, and engage designated target.
	Clear Range	Range safeties	Ensure that weapons are clear and safe.
	Three Steps Forward	All shooters and grenadiers	Echo preparatory command.
	Move	All shooters and grenadiers	Stay on line and step off three steps, counting each step aloud.
	Clear Range	Range safeties	Ensure that weapons are clear and safe.
	Seven Steps Forward	All shooters and grenadiers	Echo preparatory command.
	Move	All shooters and grenadiers	Stay on line and step off seven steps, counting each step aloud.

Figure 8-2. Sample Live-Fire Exercise (Continued)

Distance	Command	Subject	Actions
20 Meters	Shotgun, Three Point	Shotgun shooters	Load three M1013 rounds.
	M203, One Area, Two Point	M203 shooters	Load one M1026 round and have two M1006 rounds available.
	Grenadiers, Prepare to Throw	Grenadiers	Make distraction grenade ready.
	Throw	Grenadiers	Throw grenade.
		Grenadiers	Throw distraction grenade.
	Ready, Fire	Shotgun shooters	Tap down shields and engage designated targets.
		M203 shooters	Tap down shields, fire area at targets, reload with M1006 rounds, and engage designated target.
	Clear Range	Range safeties	Ensure that weapons are clear and safe.
	Shotgun, Two Point	Shotgun shooters	Load two M1013 rounds.
	M203, Two Point	M203 shooters	Load one M1006 round and have another available.
	Grenadiers, Prepare to Throw	Grenadiers	Make distraction grenade ready.
	Clear Range	Range safeties	Ensure that weapons are clear and safe.
	Three Steps Forward	All shooters and grenadiers	Echo preparatory command.
	Move	All shooters and grenadiers	Stay on line and step off three steps, counting each step aloud.

Figure 8-2. Sample Live-Fire Exercise (Continued)

Distance	Command	Subject	Actions
30 Meters	Shotgun, Two Area, Two Point	Shotgun shooters	Load two M1012 rounds and two M1013 rounds.
	M203, Two Area, Two Point	M203 shooters	Load one round M1026 and have one M1026 and two rounds M1006 available.
	Grenadiers, Prepare to Throw	Grenadiers	Make distraction grenade ready.
	Throw	Grenadiers	Throw grenade.
		Grenadiers	Throw distraction grenade.
	Ready, Fire	Shotgun shooters	Tap down shields, engage targets, and engage designated target with M1013 rounds.
		M203 shooters	Tap down shields, fire area at target, reload with M1006 rounds, and engage designated target.
	Clear Range	Range safeties	Ensure that weapons are clear and safe.
	Seven Steps Forward	All shooters and grenadiers	Echo preparatory command.
	Move	All shooters and grenadiers	Stay on line and step off seven steps, counting each step aloud.

Figure 8-2. Sample Live-Fire Exercise (Continued)

Distance	Command	Subject	Actions
40 Meters	Shotgun, Two Area, One Point	Shotgun shooters	Load two M1012 rounds and one round M1013.
	M203, One Area, One Point	M203 shooters	Load one M1026 round and have one M1006 round available.
	Grenadiers, Prepare to Throw	Grenadiers	Make distraction grenade ready.
	Throw	Grenadiers	Throw grenade.
		Grenadiers	Throw distraction grenade.
	Ready, Fire	Shotgun shooters	Tap down shields and engage targets with M1013 rounds.
		M203 shooters	Tap down shields, fire area at target, reload with M1006 rounds, and engage designated target.
	Clear Range	Range safeties	Ensure that weapons are clear and safe.
	Seven Steps Forward	All shooters and grenadiers	Echo preparatory command.
	Move	All shooters and grenadiers	Stay on line and step off seven steps, counting each step aloud.
50 Meters	Shotgun, One Area	Shotgun shooters	Load one M1012 round.
	M203, One Area, One Point	M203 shooters	Load one M1026 round and have one M1006 round available.
	Grenadiers, Prepare to Throw	Grenadiers	Make distraction grenade ready.
	Throw	Grenadiers	Throw grenade
		Grenadiers	Throw distraction grenade.
	Ready, Fire	Shotgun shooters	Tap down shields, engage targets, and engage designated target with M1013 rounds.
		M203 shooters	Tap down shields, fire area at the target, reload with M1006 rounds, and engage the designated target.
	Clear Range	Range safeties	Ensure that weapons are clear and safe.
	Seven Steps Forward	All shooters and grenadiers	Echo preparatory command.
	Move	All shooters and grenadiers	Stay on line and step off seven steps, counting each step aloud.

Figure 8-2. Sample Live-Fire Exercise (Continued)

NOTE: Shield holders are referred to as group one, shooters are referred to as group two, grenadiers are referred to as group three.

8-17. Currently, there are no qualification standards established in the Army for NL munitions. Area rounds, which contain multiple projectiles designed to affect multiple targets when engaged, only require approximately five rounds. This affords shooters an opportunity to realize the characteristics and capabilities of these rounds.

8-18. Single projectile rounds like the M1013 and the M1006 require a greater degree of proficiency. If situations call for the engagement of targets with these rounds, it is essential that trained shooters be able to hit what they are aiming at. For example, firing 10 rounds from various ranges can provide soldiers with the confidence to be effective shooters. A standard can be established that requires shooters to hit targets with at least 6 out of the 10 rounds used in the example to be considered proficient.

8-19. *Figures 8-3; 8-4,* page 8-11; *8-5,* page 8-21; and *8-6,* page 8-22, give examples of plans to follow if in a situation similar to the following scenario (see *Army Training and Evaluation Program (ARTEP) 19-313-10-MTP*):

A company commander is providing security patrols in support of a joint task force (JTF). Besides providing security patrols for a sector of responsibility, one platoon must always be designated as a quick reaction force (QRF) for civil disturbance operations. For the last 2 weeks the sector has been quiet and the commander has ordered all platoons to conduct civil disturbance training on a rotational basis.

A new order has been sent down to provide a squad-sized element to provide security for the future polling site for an upcoming election. This polling site has yet to open, but intelligence indicates potential problems as the election approaches.

On the day prior to the scheduled election, a crowd of approximately 25 to 30 demonstrators aggressively move toward the polling site, chanting slogans and yelling toward the security forces posted there. While some demonstrators begin throwing whatever they can pick up, the majority are simply yelling, locking arms, and blocking roadways. Hundreds of additional people, not part of the crowd, begin to gather to watch the activities as they unfold.

In an assessment of the demonstrators, it is determined that they intend to stay through the night into the morning of the election with the intention to block access to the polls. It is essential that the crowd is dispersed and that access to the polls is unrestricted in time for the election to begin the next morning.

Evaluation Scheme	
The evaluation will consist of five lanes conducted over a 3-day period. Platoons will execute 24-hour operations during the evaluation. Any retraining will be done immediately following the execution of the lane if time permits. The following tasks will be evaluated:	
Lane 1	Cordon and search operations
Lane 2	Civil disturbance QRF
Lane 3	Security and defense of a critical site (polling site)
Lane 4	Night civil disturbance QRF
Lane 5	Night LP/OP (PIO)

Evaluation Schedule				
Date and Time	**1st Platoon**	**2d Platoon**	**3d Platoon**	**4th Platoon**
	Lane 1	Lane 2	Lane 3	Lane 4
	Lane 2	Lane 3	Lane 4	Lane 5
	Lane 3	Lane 4	Lane 5	Lane 1
	Lane 4	Lane 5	Lane 1	Lane 2
	Lane 5	Lane 1	Lane 2	Lane 3

Figure 8-3. Evaluation Scheme

Unit Training Schedule and Firing Tables

Unit Training Schedule 1

Training Day 1

Time	Subject	Location	Uniform	Instructor	Notes
0800	Administrative time	Classroom	U1		
0830	GRM	Classroom	U1		
1000	ROE and use of force	Classroom	U1		
1200	Chow				
1300	MOOTW	Classroom	U1		
1500	Effective communication	Classroom	U1		

Administration:

Uniforms:

U1: BDU

U2: BDU and LBE with harness and canteens

U3: BDU and LBE with harness, canteens, flak jacket, and Kevlar helmet

U4: BDU and LBE with harness, canteens, flak jacket, Kevlar helmet, and assigned weapon

Transportation:

Weapons:

Ammunition:

Required Equipment for Training:

1. Projection screen
2. Video projector with CPU
3. Television
4. VCR

Miscellaneous:

1. All personnel are required to have a notebook and a pencil or pen.
2. Handouts must be prepared and passed out.

Figure 8-4. Sample Two-Week Plan

Training Day 2

Time	Subject	Location	Uniform	Instructor	Notes
0800	NLCS	Classroom	U1		
0900	PPCT and empty hand	Classroom	U1		
1000	PPCT and empty hand (Practical application)	PT field	U2		
1200	Chow				
1300	PPCT and empty hand (Practical application)	PT field	U2		
1700	COB				

Administration:

Uniforms:

U1: BDU

U2: BDU and LBE with harness and canteens

U3: BDU and LBE with harness, canteens, flak jacket, and Kevlar helmet

U4: BDU and LBE with harness, canteens, flak jacket, Kevlar helmet, and assigned weapon

Transportation:

Weapons:

Ammunition:

Required Equipment for Training:

1. Projection screen
2. Video projector with CPU
3. Training strike bags
4. Flex-Cufs

Miscellaneous:

1. All personnel are required to have a notebook and a pencil or pen.
2. Handouts must be prepared and passed out.
3. Medical personnel must be on location.

Figure 8-4. Sample Two-Week Plan (Continued)

Training Day 3

Time	Subject	Location	Uniform	Instructor	Notes
0800	PPCT and empty hand (Practical application)	PT field	U2		
1200	Chow				
1300	PPCT and empty hand (Practical application)	PT field	U2		
1700	COB				

Administration:

Uniforms:

U1: BDU

U2: BDU and LBE with harness and canteens

U3: BDU and LBE with harness, canteens, flak jacket, and Kevlar helmet

U4: BDU and LBE with harness, canteens, flak jacket, Kevlar helmet, and assigned weapon

Transportation:

Weapons:

Ammunition:

Required Equipment for Training:

1. Projection screen
2. Video projector with CPU
3. Training strike bags
4. Flex-Cufs

Miscellaneous:

1. All personnel are required to have a notebook and a pencil or pen.
2. Handouts must be prepared and passed out.
3. Medical personnel must be on location.

Training Day 4

Time	Subject	Location	Uniform	Instructor	Notes
0800	Riot baton training	Classroom	U1		
0900	Riot baton training (Practical application)	PT field	U2		
1200	Chow				

Figure 8-4. Sample Two-Week Plan (Continued)

1300	Riot baton training (Practical application)	PT field	U2		
1700	COB				

Administration:

Uniforms:

U1: BDU

U2: BDU and LBE with harness and canteens

U3: BDU and LBE with harness, canteens, flak jacket, and Kevlar helmet

U4: BDU and LBE with harness, canteens, flak jacket, Kevlar helmet, and assigned weapon

Transportation:

Weapons:

1. Expandable riot baton with carrier
2. Training riot batons

Ammunition:

Required Equipment for Training:

1. Projection screen
2. Video projector with CPU
3. Training strike bags
4. Flex-Cufs

Miscellaneous:

1. All personnel are required to have a notebook and a pencil or pen.
2. Handouts must be prepared and passed out.
3. Medical personnel must be on location.

Training Day 5

Time	Subject	Location	Uniform	Instructor	Notes
0800	Riot baton training (Practical application)	PT field	U2		
1000	Crowd dynamics	Classroom	U1		
1130	Chow				
1230	Crowd dynamics	Classroom	U1		
1330	CD formations	Classroom	U1		

Figure 8-4. Sample Two-Week Plan (Continued)

1430	RCAs and OC	Classroom	U1
1700	COB		

Administration:

Uniforms:

U1: BDU

U2: BDU and LBE with harness and canteens

U3: BDU and LBE with harness, canteens, flak jacket, and Kevlar helmet

U4: BDU and LBE with harness, canteens, flak jacket, Kevlar helmet, and assigned weapon

Transportation:

Weapons:

1. Expandable riot baton with carrier
2. Training riot baton

Ammunition:

Required Equipment for Training:

1. Projection screen
2. Video projector with CPU
3. Training strike bags
4. Flex-Cufs

Miscellaneous:

1. All personnel are required to have a notebook and a pencil or pen.
2. Handouts must be prepared and passed out.
3. Medical personnel must be on location.

Training Day 6

Time	Subject	Location	Uniform	Instructor	Notes
0800	RCAs and OC (Practical application)	PT field	U2		
1200	Chow				
1300	RCAs and OC (Practical application)	PT field	U2		
1700	COB				

Figure 8-4. Sample Two-Week Plan (Continued)

Administration:

Uniforms:

U1: BDU

U2: BDU and LBE with harness and canteens

U3: BDU and LBE with harness, canteens, flak jacket, and Kevlar helmet

U4: BDU and LBE with harness, canteens, flak jacket, Kevlar helmet, and assigned weapon

Transportation:

Weapons:

Ammunition:

1. MK inert OC
2. MK 4 live OC disperser
3. M36 live CR disperser

Required Equipment for Training:

1. Five large clean trash cans with liners
2. Two water hoses with running water

Miscellaneous:

Medical personnel must be on location.

Training Day 7

Time	Subject	Location	Uniform	Instructor	Notes
0800	NL munitions employment	Classroom	U1		
1030	Chow				
1115	Safety brief	Range	U4		
1130	NL munitions (live fire)	Range	U4		
TBD	Evening chow				
TBD	Safety brief	Range	U4		
TBD	NL munitions night fire exercise	Range	U4		
TBD	Clean up and COB				

Figure 8-4. Sample Two-Week Plan (Continued)

Administration:

Uniforms:

U1: BDU

U2: BDU and LBE with harness and canteens

U3: BDU and LBE with harness, canteens, flak jacket, and Kevlar helmet

U4: BDU and LBE with harness, canteens, flak jacket, Kevlar helmet, and assigned weapon

Transportation:

Yes

Weapons:

1. M16A2 with M203
2. 12-gauge shotgun

Ammunition:

1. Fin-stabilized M1012
2. M1013 multipellet (rubber)
3. M1006 40-millimeter sponge grenade
4. M1026 40-millimeter multiball (rubber)
5. M84 diversionary grenade

Required Equipment for Training:

1. Fifteen to twenty human form targets
2. Eye and hearing protection
3. Fifteen to twenty riot shields
4. Two MREs per student
5. Video recorder
6. Three to five high-intensity lights
7. All students should be issued face shields, shin guards, and ammunition pouches

Miscellaneous:

1. Medical personnel must be on location.
2. Range safety officers must be available.
3. Primary and secondary means of communication with range control must be available.
4. Operational risk assessments must be conducted.

Figure 8-4. Sample Two-Week Plan (Continued)

Training Day 8

Time	Subject	Location	Uniform	Instructor	Notes
0800	CD formations (Practical application)	TA 1	U4		
1200	Chow				
1300	CD formations (Practical application)	TA 1	U4		
1700	COB				

Administration:

Uniforms:

U1: BDU

U2: BDU and LBE with harness and canteens

U3: BDU and LBE with harness, canteens, flak jacket, and Kevlar helmet

U4: BDU and LBE with harness, canteens, flak jacket, Kevlar helmet, and assigned weapon

Transportation:

Yes

Weapons:

1. Training weapons
2. Training riot batons

Ammunition:

Required Equipment for Training:

1. Eye protection
2. Fifteen to twenty riot shields
3. Video recorder
4. Three to five high-intensity lights
5. Face shields, shin guards, and ammunition pouches

Miscellaneous:

1. Medical personnel must be on location.
2. Primary and secondary means of communication must be available.
3. Operational risk assessment must be conducted.
4. Twenty to thirty OPFOR personnel must be available.

Figure 8-4. Sample Two-Week Plan (Continued)

Training Day 9

Time	Subject	Location	Uniform	Instructor	Notes
0800	Introduction to M5, PVAB, and M315	Classroom 1	U1		
1200	Chow				
1300	M5, PVAB, and M315 (Practical application/LFE)	Range	U4		
1700	COB				

Administration:

Uniforms:

U1: BDU

U2: BDU and LBE with harness and canteens

U3: BDU and LBE with harness, canteens, flak jacket, and Kevlar helmet

U4: BDU and LBE with harness, canteens, flak jacket, Kevlar helmet, and assigned weapon

Transportation:

Yes

Weapons:

Two complete M315s and installation kits

Ammunition:

1. Ten live MCCMs
2. Ten live M315 rounds of each

Required Equipment for Training:

1. Eye protection
2. Ten inert MCCMs
3. Two complete PVAB systems

Miscellaneous:

1. Medical personnel must be on location.
2. Primary and secondary means of communication must be available.
3. Operational risk assessment must be conducted.

Figure 8-4. Sample Two-Week Plan (Continued)

Training Day 10					
Time	**Subject**	**Location**	**Uniform**	**Instructor**	**Notes**
0800	Tactical consideration	MOUT classroom	U1		
0930	Safety brief	MOUT classroom	U4		
1000	Tactical exercise	MOUT Facility	U4		

Administration:

Uniforms:

U1: BDU

U2: BDU and LBE with harness and canteens

U3: BDU and LBE with harness, canteens, flak jacket, and Kevlar helmet

U4: BDU and LBE with harness, canteens, flak jacket, Kevlar helmet, and assigned weapon

Transportation:

Yes

Weapons:

1. Assigned weapons
2. Inert RCA/OC individual dispersers
3. Training riot batons
4. Fifty practice grenades

Ammunition:

Required Equipment for Training:

1. Eye protection
2. Ten inert MCCMs
3. Face shields, shin guards, and ammunition pouches
4. Fifteen to twenty full-length riot shields

Miscellaneous:

1. Medical personnel must be on location.
2. Primary and secondary means of communication must be available.
3. Operational risk assessment must be conducted.
4. A video recorder must be available.

Figure 8-4. Sample Two-Week Plan (Continued)

Sample Unit Training Schedule				
Training Day 1				
Time	**Event**	**Location**	**Instructor**	**Notes**
0730-0830	Force continuum	Classroom		L
0830-0930	Management of aggressive behavior	Classroom		L
0930-1030	ROE	Classroom		L
1030-1130	Crowd control	Classroom		L
1130-1230	Chow			
1230-1700	PPCT and empty hand	PT field		L, PA
Training Day 2				
Time	**Event**	**Location**	**Instructor**	**Notes**
0730-1130	Riot baton training	PT field		L, PA
1130-1230	Chow			
1230-1530	Riot baton training	PT field		PA
1530-1700	CD formations	Classroom		L
Training Day 3				
Time	**Event**	**Location**	**Instructor**	**Notes**
0730-1130	CD formations	PT field		PA
1130-1230	Chow			
1230-1700	NL munitions, NLW	Range		L, PA
Training Day 4				
Time	**Event**	**Location**	**Instructor**	**Notes**
0730-1130	RCAs and OC	Classroom and PT field		L, PA
1130-1230	Chow			
1230-1700	RCAs and OC	PT field		PA

Figure 8-5. Sample One-Week Plan

Training Day 5				
Time	**Event**	**Location**	**Instructor**	**Notes**
0730-1700	Tactical exercise	MOUT facility		OPS
NOTE: Appointed student leaders ensure that troops are fed and rotated for rest.				

Figure 8-5. Sample One-Week Plan (Continued)

Sample Unit Training Schedule				
Training Day 1				
Time	**Event**	**Location**	**Instructor**	**Notes**
	Shield and riot baton techniques	MOUT site		Individual and collective tasks
	NLCS and munitions	Range Classroom		Leader training
	Formation training	MOUT site		Individual and collective tasks
	Chow			
	Formation training	MOUT site		Individual and collective tasks
Training Day 2				
Time	**Event**	**Location**	**Instructor**	**Notes**
0730-1930, 12-hour operations	Live-fire exercise	Range		M203/12-gauge
	QRF exercise	MOUT site		Individual and collective tasks
NOTE: Once weapons and munitions familiarization is complete, shooters join the QRF exercise to be used. For live-fire range setup, see *Figure 8-1.*				

Figure 8-6. Sample Two-Day Plan

Appendix A

Metric Conversion Chart

This appendix complies with current Army directives which state that the metric system will be incorporated into all new publications. *Table A-1* is a metric conversion chart.

Table A-1. Metric Conversion Chart

US Units	Multiplied By	Metric Units
Cubic feet	0.0283	Cubic meters
Feet	0.3048	Meters
Feet per second	0.3048	Meters per second
Inches	0.0254	Meters
Inches	25.4001	Millimeters
Miles per hour	1.6093	Kilometers per hour
Pounds	0.4536	Kilograms
Yards	0.9144	Meters
Metric Units	**Multiplied By**	**US Units**
Cubic meters	35.3146	Cubic feet
Kilograms	2.2046	Pounds
Kilometers per hour	0.6214	Miles per hour
Meters	3.2808	Feet
Meters	39.3701	Inches
Meters	1.0936	Yards
Meters per second	3.2808	Feet per second
Millimeters	0.0394	Inches

Appendix B

Operations and Legal Considerations in the Continental United States

Providing military assistance to state and local governments to assist them in quelling a civil disturbance or riot requires close coordination through a host of state and federal agencies. It requires a thorough briefing of soldiers at all levels on what they can and cannot do with respect to law enforcement. Civil authorities must be briefed on the restrictions placed on federal forces by the *Constitution of the United States* and federal statutes and laws. The national guard (NG), when not in a federal status, operates under the control of the state governor and the adjutant general (AG). The NG has historically been the first military responder during emergencies.

FEDERAL INTERVENTION AND AID

B-1. Under the *Constitution of the United States* and *United States Code*, the President is empowered to direct federal intervention in civil disturbances to—

- Respond to state requests for aid in restoring order.
- Enforce the laws of the United States.
- Protect the civil rights of citizens.
- Protect federal property and functions.

B-2. Under the *Constitution of the United States*, each state is responsible for protecting life and property within its boundaries. State and local governments use their LEA forces to maintain law and order and to quell civil disturbances. However, if civil disturbance support requirements exceed the resources of a state, federal troops may be requested to help restore and maintain law and order.

B-3. The *Constitution of the United States* and federal statutes authorize the President to direct the use of federal armed troops within the 50 states and territories and their political subdivisions. The President is also empowered to federalize the NG of any state to suppress rebellion and enforce federal laws.

B-4. Federal assistance is provided to a state when the state has used all of its resources, including its NG, to quell a disorder and finds the resources insufficient. Usually, active duty federal forces are used to augment the NG of the requesting state. However, the President may choose to federalize the NG of another state and use them alone or with other forces to restore order.

B-5. The President may employ federal troops to ensure the execution of US law when a state opposes or obstructs US law or impedes the course of justice under those laws. The President may employ armed federal troops to suppress insurrection, domestic violence, unlawful assemblies, and conspiracy if such acts deprive the people of their constitutional rights and privileges and the civil authorities of a state cannot or will not provide adequate protection.

B-6. The president is also authorized to use armed federal troops to protect federal property and functions when the need for protection exists and local civil authorities cannot or will not give adequate protection. The right of the United States to protect all federal property and functions regardless of their locations is an accepted principle of our government.

B-7. As a temporary measure, federal military equipment and facilities may be loaned to law enforcement agencies of state and local governmental bodies for use during civil disturbances. These resources may also be loaned to state NG and non-DOD federal agencies. The requesting agencies are expected to provide enough resources of their own to minimize the need for US military resources. In addition, the loan of the resources must not conflict with US military needs.

ROLES AND RESPONSIBILITIES OF VARIOUS AGENCIES

B-8. Several agencies within the federal and state governments play a key role in supporting the request of local authorities for support during a civil disturbance.

SECRETARY OF DEFENSE

B-9. The Secretary of Defense retains approval for federal support to civil authorities involving the use of DOD forces, personnel, and equipment. Approval may also involve DOD support that will result in a planned event with the potential for confrontation with specifically identified individuals and groups or will result in the use of lethal force.

SECRETARY OF THE ARMY

B-10. The Secretary of the Army, as an executive agent for the Secretary of Defense, is the approval authority for federal emergency support in response to natural and man-made disasters (except weapons of mass destruction [WMD]). Military forces respond to directions and guidance from the director of military support (DOMS) (the action agent for the Secretary of the Army).

DIRECTOR OF MILITARY SUPPORT

B-11. The DOMS plans and executes the DOD domestic support mission to civil authorities. He is a general officer appointed by the Secretary of the Army. The DOMS is the DOD primary contact for all federal departments and agencies during DOD involvement in most domestic support operations.

STATE COORDINATING OFFICER

B-12. The governor appoints the state coordinating officer (SCO) to coordinate disaster operations for the state. The SCO is the primary point of contact for the federal coordinating officer (FCO) in facilitating disaster assistance.

DIRECTOR OF THE FEDERAL EMERGENCY MANAGEMENT AGENCY

B-13. The director of the Federal Emergency Management Agency (FEMA), on behalf of the President, appoints the FCO. The FCO coordinates the timely delivery of federal disaster assistance to the affected state and local governments and disaster victims. He works closely with the SCO to determine state requirements and to coordinate national level issues with the catastrophic disaster response group. There must be close and continuous coordination with FEMA to ensure the proper use of military assets in support of civil authorities.

DEFENSE COORDINATING OFFICER

B-14. The defense coordinating officer (DCO) operates under the authority of the President and is the DOD representative designated to coordinate on-the-scene activities with the FCO. FEMA and other federal agencies that request support from DOD go through the DCO for validation and resourcing.

NATIONAL GUARD

B-15. The NG, in state status (see *Title 32, United States Code {USC}),* is the governor's primary military response organization for emergencies and disasters. The NG responds and operates under state laws unless they are federalized, when they operate under the same constraints and laws as federal units.

UNITED STATES ARMY RESERVE

B-16. The US Army Reserve is capable of providing military support. This assistance and support may include the use of equipment and other resources, including units and individuals.

STATE AND LOCAL GOVERNMENT RESPONSIBILITIES

B-17. State and local government officials, operating under the authority granted by state constitutions and the *Constitution of the United States,* are responsible for daily safety and security issues that impact the quality of life of their citizens. State and local officials have primary responsibility for emergency preparedness.

STATE RESPONSIBILITIES

B-18. Many state government agencies are responsible for coming to the aid of state citizens.

Governor

B-19. A state governor is empowered by each state constitution and the *Constitution of the United States* to execute the laws of the state and to command the state NG while it is serving in state status. Governors are also responsible for issuing executive orders or emergency proclamations, such as declaring states of emergency and ensuring that state agencies plan for potential disasters and civil disorder.

B-20. Once a man-made or natural disaster occurs, the governor assesses its extent and determines if local government requests for assistance should be honored. If appropriate, the governor declares a state of emergency, activates the state emergency response plan, and may order the NG into state active duty. The governor assigns the NG its mission through the office of emergency service (OES)/emergency medical dispatching (EMD).

Plans, Operations, and Military Support Officer

B-21. Each state NG has a plans, operations, and military support officer (POMSO) who prepares and coordinates contingency plans for military support to civil authorities (MSCA) during disaster response and recovery operations. The POMSO coordinates training plans and exercises between the state NG and federal, state, and local emergency management agencies. The POMSO serves as the NG point of contact with DOD officials during a declared federal emergency or disaster.

Office of Emergency Services

B-22. Each state has a specific agency and/or office that coordinates emergency preparedness and serves as the governor's coordinating agency during an emergency. The titles of these offices vary from state to state. Some examples are the Division of Emergency Government, Emergency Management Agency, Department of Public Safety, or Office of Emergency Preparedness. This manual refers to this office using the generic term OES/EMD.

B-23. Generally, the OES/EMD is organized as a stand-alone office under the governor or aligned under the AG or the state police, and in some instances, the AG is the OES/EMD director. It operates the state emergency operations center and coordinates with state agencies and federal officials for assistance.

Adjutant General

B-24. The state NG is the governor's primary military response force in an emergency. The AG, through the state area command (STARC) (specifically the POMSO), develops and coordinates emergency response plans for disasters and emergencies. The AG is in command of state NG forces ordered into state active duty.

State Government Agencies

B-25. State government departments and agencies prepare emergency preparedness plans for their areas of specialization in support of the governor's plan. They also participate in emergency preparedness exercises.

LOCAL RESPONSIBILITIES

B-26. Mayors, city managers, local police, fire protection officials, county executives, sheriffs, prosecuting attorneys, and public health officials are some of the people responsible for law enforcement, safety, health, and fire protection on a daily basis. They are responsible for developing appropriate emergency response plans and responding to emergencies within their jurisdictions. Most local jurisdictions have an OES/EMD to plan and coordinate actions in an emergency. In many cases, local jurisdictions have mutual aid agreements with other jurisdictions that allow for firefighting, medical, and police assistance across jurisdictional boundaries. Once local officials determine that an emergency is beyond the scope of their resources or ability to respond, the senior local official is responsible for requesting additional assistance from the next agency in the assistance request process.

LEGAL CONSIDERATIONS AND CONSTRAINTS

B-27. The *Constitution of the United States*, laws, regulations, policies, and other legal issues limit the use of federal military personnel in domestic support operations. Any Army involvement in civil disturbance operations involves many legal issues requiring comprehensive legal reviews.

B-28. Under the *Constitution of the United States*, Congress has the authority to raise and support an army, provide and maintain a navy, and make rules for governing and regulating the land and naval forces. The *Constitution of the United States* places the military under civilian control and designates the President as commander-in-chief. Statutes provide for civilian leadership in the office of the secretary of defense, service secretaries, and various other civil authorities. Under the *Constitution of the United States*, the civilian government must preserve public order and carry out governmental operations within its territorial limits. The *Constitution of the United States* allows the use of the federal military to execute or enforce the law when necessary to protect federal or civilian property and functions. However, significant restrictions exist on employing federal military forces within the United States.

B-29. The unique capabilities of the military enable it to support federal, state, or local civilian agencies. In most circumstances, the DOD is one of many federal agencies reacting to a domestic emergency or crisis, playing a subordinate, supporting role to a lead civilian agency.

LAW

B-30. In the United States, civilian agencies provide for the needs of citizens. Civilian, federal, state, and local government and law enforcement agencies execute US laws. Laws governing the use of the military in domestic operations are complex, subtle, and constantly changing. Commanders involved in civil disturbance operations should staff plans, policies, programs, exercises, funding, operations, constraints, and limitations with their local SJA to ensure conformity with legal requirements. They should scrutinize each request for aid (whether it is for equipment, individuals, or training) to ensure that it conforms to statutory requirements. The following references are a snapshot of the law on military support to civilian authorities:

- *Section 331, Chapter 15, Part 1, Subtitle A, Title 10 USC.*
- *Section 332, Chapter 15, Part 1, Subtitle A, Title 10 USC.*
- *Section 333, Chapter 15, Part 1, Subtitle A, Title 10 USC.*
- *Section 334, Chapter 15, Part 1, Subtitle A, Title 10 USC.*
- *Section 1385, Chapter 67, Part 1, Title 18 USC.*
- *DOD Directive 3025.12.*

- *DOD Directive 3025.15*
- *DOD Directive 5525.5*
- *DOD Civil Disturbance Plan 55-2*
- *NGR 500-1/ANGI 10-8101*
- *FM 3-19.15*

INFORMATION OPERATIONS

B-31. Information superiority helps forces anticipate problems and requirements. It allows commanders to control situations earlier and with less force, creating the conditions necessary to achieve the end state. Public affairs, PSYOP, and civil military operations are activities that will allow the commander to control situations earlier and with less force.

B-32. Information is available from a multitude of sources, the primary being open sources, law enforcement, and the military. A diversity of sources is the best approach because it prevents biased behavior.

Open sources include—

- Libraries.
- Newspapers and news periodicals.
- Radio and television.

Law enforcement sources include—

- Local law enforcement agencies.
- National law enforcement agencies.

Military sources include—

- DOD intelligence community (most restrictive source).
- Local MI field offices.

B-33. The DOD intelligence community operates under limitations imposed by regulations and executive orders. Attempts to skirt these restrictions may violate regulations or federal statutes.

RESTRICTIONS

B-34. Strict limits are placed on collecting information related to a civil disturbance in order to protect the civil rights of people and organizations not affiliated with the DOD. Civil disturbance plans and materials must not include lists of groups or people not affiliated with the DOD. Information on civilians and civilian organizations can be collected only with specific authorization from the Secretary or the Under Secretary of the Army. Conditions for collecting information include the existence of threats against Army personnel, functions, or property (see *ARs 380-10* and *380-13).* Civil disturbance information (available in public documents) or open-source information may be collected. However, specific rules regarding its storage must be followed. Commanders must coordinate with SJA, MI, and USACIDC before collecting any such information.

B-35. Civilian law enforcement agencies may limit access to law enforcement intelligence. In such circumstances, the military may request that agencies release the unclassified version.

B-36. If the Department of Justice (DOJ) determines that federal intervention into a civil disturbance is likely, then information relating to the disturbance is provided to the Army Assistant Chief of Staff for Intelligence. The information is analyzed and then passed through channels to the TF commander for planning purposes.

B-37. Military intelligence collection efforts, except liaison, may begin only when DA orders. The orders will come through the CSA's personal liaison officer and the TF commander. Covert operations to gather information on non-DOD individuals and groups must be approved on an operation-by-operation basis by the Undersecretary of the Army.

B-38. Once DA approves the collection effort, MI elements establish and maintain liaison with the appropriate local, state, and federal authorities. Then they collect information pertaining to the incident and general situation. Their estimate of the situation is passed to the TF commander and DA.

POSSE COMITATUS

B-39. Generally, federal military forces may not give law enforcement assistance to civil authorities without conflicting with the *Posse Comitatus Act*. However, constitutional and statutory exceptions to this prohibition do exist. The recent emphasis on drug interdiction has led to an increase in those exceptions.

B-40. The *Judiciary Act of 1789* allowed US marshals to call upon the military as a posse comitatus. This continued until after the Civil War, when the federal government used the Army to execute reconstruction era policies. The southern states regarded the use of the military for this purpose as abusive and repressive, and in 1878, President Rutherford B. Hayes signed the original bill ending the practice.

B-41. The *Posse Comitatus Act* prescribes criminal penalties for the use of the US Army or Air Force to execute the laws of or to perform civilian law enforcement functions within the United States. DOD policy extends this prohibition to the US Navy and Marine Corps. Prohibiting the military from executing the laws means that military personnel may not participate directly in—

- An arrest; a search and seizure; a stop and frisk; or an interdiction of vessels, aircraft, or vehicles.
- A surveillance or pursuit.
- A civilian legal case or any other civilian law enforcement activity as informants, undercover agents, or investigators.

B-42. The *Posse Comitatus Act* does not apply to the following:

- Members of the NG when in state service.
- Members of a Reserve Component when not on active duty or active duty for training.
- DOD civilians, unless under the direct command and control of an active duty officer.

- A soldier when off duty and acting only in a private capacity.
- Soldiers taking action for the primary purpose of furthering a military or foreign affairs function of the United States; for example, enforcing military justice, maintaining law and order on military installations, and protecting classified materials.

CONSTITUTIONAL EXCEPTIONS TO THE POSSE COMITATUS ACT

B-43. Under its inherent authority, the US government is responsible for preserving public order and carrying out governmental operations within its territorial limits, by force if necessary. Under the *Constitution of the United States*, exceptions allow the use of military force to execute or enforce the law. Some of the exceptions are to—

- Protect civilian property and functions.
- Ensure preservation of public order and carry out government operations, by force if necessary.
- Protect civil rights where local authorities do not or cannot protect them.
- Protect federal property and functions.
- Quell civil disturbances and labor strife that rises to the level of civil disorder.

B-44. The President may order the armed forces to support state civil authorities suffering from an insurrection or civil disturbance. He must act personally by first issuing a proclamation calling on insurgents to disperse and retire peacefully within a limited time. This occurred when federal forces were called in to counter the Los Angeles riots in 1992.

Protect Civilian Property and Functions

B-45. A sudden and unexpected civil disturbance, disaster, or calamity may seriously endanger life and property and disrupt normal governmental functions to such an extent that local authorities cannot control the situation. At such times, the federal government may use military force to prevent loss of life or wanton destruction of property and to restore government functions and public order. This exception has rarely been used.

Protect Federal Property and Functions

B-46. The federal government may use military force to protect federal property and federal government functions when local authorities cannot or decline to provide adequate protection.

ARMED FORCES

B-47. Basic guidance for the interaction of the military, the Reserve Component, and civilian LEAs is given in *10 USC Chapter 18.* Guidelines on reimbursement and restrictions, direct participation in law enforcement activities, the use of information collected during military operations, and the use of military equipment and facilities are some of the topics covered. Title 10 prohibits the military from directly participating in arrests, searches, seizures, or other similar activity unless authorized by law (such as arrests on military property). The *Fiscal Year 1989 Act* and subsequent *National Defense Authorization Acts* have authorized the DOD to provide more support to LEAs in the counterdrug effort.

NATIONAL GUARD

B-48. For a description of how the secretary of defense may provide funds to state governments (including the District of Columbia, Commonwealth of Puerto Rico, and US territories) for civil disturbance operations performed by the National Guard when not in federal service, see 32 *USC*. The *Posse Comitatus Act* does not apply to the NG when not in federal service. Each state specifies how NG may be employed.

SUPPORT TO CIVIL LAW ENFORCEMENT

B-49. It is DOD policy to cooperate with civilian law enforcement officials to the greatest practical extent. However, cooperation must consistently meet the needs of national security and military preparedness, the historic tradition of limiting direct military involvement in civilian LEAs, and the requirements of applicable law. US military forces are never placed under the command of civilian law enforcement officers or nonfederalized NG. DOD policy concerning the provision of military support to LEAs, including personnel and equipment, are contained in *DOD Directive 5525.5*.

B-50. The Army assists civilian law enforcement by providing personnel, equipment, training, and expert advice within the limits of applicable law. Army NG units in state status (see 32 *USC*) provide the primary source of military assistance. They may be able to provide assistance to civil authorities in instances when federal units are precluded due to the restrictions of the *Posse Comitatus Act*.

ROLES OF THE NATIONAL GUARD AND FEDERAL FORCES

B-51. The preservation of law and order in the civilian community is the responsibility of state and local governments and law enforcement authorities. When a civil disturbance in a civilian community turns to widespread rioting that includes arson, looting, and acts of violence, the civil authorities may decide that they do not have the resources to quell the riot. They may then turn to the NG and federal forces to support the civil authorities in restoring law and order. Civil disturbances in any form are prejudicial to public law and order.

NATIONAL GUARD FORCES

B-52. The NG (as a state organization) responds to the governor according to state law for civil disturbance operations. NG regulations direct planning and training for the civil disturbance mission. During most civil disturbance situations, the NG will be the first military responder and will usually remain in state active duty status (according to 32 *USC*) throughout the operation. In extreme circumstances, the NG can be brought on federal service for civil disturbance operations when ordered to under the appropriate federal statute by the President. During the 1992 Los Angeles civil disturbance, the President federalized the California NG. Once the NG has been deployed to assist the civil authorities with a civil disturbance, it is too far into the deployment cycle to start determining what tasks the NG can do as a state or as a federally controlled force. It is the commander's responsibility not to accept missions that the unit is not adequately trained to do. The following are examples of appropriate missions for the NG:

- Manning traffic control points.
- Providing building security.
- Providing area security and patrols.
- Providing security at custody facilities.
- Providing security and escort for emergency personnel and equipment.
- Protecting sensitive sites.
- Transporting law enforcement personnel.
- Providing show of force.
- Dispersing crowds.
- Employing riot control munitions.
- Providing very important person (VIP) protection and escort.
- Providing quick reaction and reserve force.

B-53. Examples of inappropriate taskings:

- Negotiating hostage situations.
- Barricading suspects.
- Participating in criminal investigations.
- Acting as federal Army forces.

B-54. The commitment of federal troops to deal with domestic civil disturbances must be viewed as a drastic measure of last resort. Their role should never be greater than what is absolutely necessary under the circumstances. Commanders should take every measure to avoid the perception of an invading force. A JTF designated to respond to a civil disturbance should project the image of a restrained and well-disciplined force whose sole purpose is to help restore law and order with minimal harm to the people and property and with due respect for all law-abiding citizens.

B-55. The role of federal Army forces is to assist civil authorities in restoring law and order when the magnitude of the disturbance exceeds the capabilities of local and state LEA, including the NG. Under the provisions of the *Constitution of the United States* and selected federal statutes, the President may order the employment of the federal armed forces to aid local and state civil authorities to protect the Constitutional rights of citizens. Federal military forces may also protect federal facilities and installations in any state, territory, or possession.

B-56. The *DA Civil Disturbance Plan 55-2* gives guidance for Army forces directed to quell civil disturbances. In addition, *AR 500-50* outlines responsibilities, prescribes policies, and provides guidance for the use of Army resources in response to civil disturbances. It outlines statutory restrictions and designates staff and command responsibilities for planning and executing civil disturbance operations.

B-57. The JTF commander exercises control of all federal military forces (including NG in federal status) committed to assist civil authorities. Federal military forces remain under the military chain of command during civil disturbance operations. Federal forces will not be placed under the command of either civil officials or NG commanders in a nonfederal status. Civil authorities retain control of their state and local law enforcement agencies. The JTF commander establishes liaison with the senior representative of the attorney general (SCRAG) and other appropriate federal, state, and local civil authorities.

B-58. When a JTF is being assembled it must deploy with a provost marshal (PM) who will be the POC for situation assessment, operational information, and security matters. He also serves as a liaison with civilian and law enforcement agencies.

B-59. Federal military forces must be tailored to the specific civil disturbance situation. Sufficient combat support and combat service support units will be required to sustain the force throughout the deployment. Coordination with civil officials may allow the force to draw on resources available from state and local agencies. Close and continuous coordination between federal military forces and LEAs will provide the commander with the detailed and timely information required to employ and protect the force effectively.

B-60. In supporting *DA Civil Disturbance Plan 55-2,* intelligence personnel may conduct close and continuous liaison with LEAs and military police to ensure that their units receive the information needed to allow the commander to adequately protect the force. The JTF commander should staff intelligence support missions with his senior intelligence officer and legal counsel before approving the mission.

B-61. Federal military forces must be employed in tasks or missions appropriate to their organization and training; they must not be employed in ways that violate the legal restrictions in effect. Military forces may be used to disperse unlawful assemblies and to patrol disturbed areas to prevent unlawful acts. They may be used to assist in the distribution of essential goods and the maintenance of essential services. Forces may also establish traffic control points, cordon off areas, release smoke and obscurants, and serve as security or quick reaction forces. Certain types of missions (such as gathering intelligence on civilians) are always inappropriate for military forces during civil disturbance operations.

B-62. Requests for specific military missions are typically passed through a single state or federal law enforcement coordinating officer, as approved by the SCRAG. Validated requests are transmitted to the JTF commander and his HQ for staffing and coordination. Approved missions are assigned through the military chain of command to the appropriate element or unit for execution. Units and soldiers will not accept taskings or missions directly from law enforcement or civilian officials, except in a direct support relationship as approved and ordered through the military chain of command.

B-63. It is important that a military liaison be provided to each LEA HQ generating requests for support. This liaison can assist LEA officials in determining the types and quantities of military support to be requested. JTF HQ can facilitate this mission assignment process by providing LEAs with a detailed listing of the types of missions military forces may conduct.

B-64. The AO for a deployed unit should coincide with the jurisdiction or subdivision boundaries of the LEA it supports. This arrangement facilitates liaison and coordination between law enforcement and military chains of command.

LEAD AGENCY CONCEPT AND ROLE OF MILITARY

B-65. The DOJ is the lead federal agency for civil disturbance operations. The Attorney General's on-scene representative is the SCRAG. The DOD has designated DA as their executive agent for military assistance for civil disturbance (MACDIS) (see *FM 27-100.*)

JOINT TASK FORCE

B-66. Military support to civil authorities in disasters and domestic emergencies is a DOD (not a service component) responsibility. The supported commander may decide that the severity and scope of a disaster require a joint response. In civil assistance, as in operational level warfighting, the commander-in-chief (CINC) uses the different and complementary capabilities of each service to accomplish the mission. Such use requires knowledge of both the capability and the availability of all service component assets, to include their agencies and installation.

B-67. The CINC may establish a JTF to provide comprehensive military support. Establishing a JTF may provide the best DOD response to civil emergencies. A JTF is established to execute a specific mission limited in scope and duration. The JTF objective in civil emergencies is to deploy forces to the area rapidly to assist immediately in saving lives and safeguarding property and to continue providing assistance required by the federal response plan (FRP) and the particular situation.

B-68. The JTF is configured for each specific mission. In some civil emergencies, the JTF may require a greater proportion of combat service support type units and capabilities than in typical warfighting deployments. The JTF must be able to provide emergency assistance across all lines of support.

JOINT TASK FORCE CONTROL ELEMENT

B-69. Responsibility for controlling the civil disturbance resides with the civil authorities. The JTF is organized to support and lend assistance to them in their effort to return civil order to the community.

B-70. The control element consists of the command group and the crisis management team (CMT). The command group is made up of city, county, state, and military command personnel. They set policy and issue directives, coordinate the activities of civil and military support agencies, supervise the CMT, and coordinate with outside agencies. The CMT is made up of representatives of civilian and military staff sections. The CMT advises the command group and coordinates operations and support for the action element of the TF. The control elements are located with the emergency operations center (EOC) to facilitate information processing, resource management, and operational control. The JTF command element should colocate with the civilian command element at the EOC. This will expedite the exchange of information. If an EOC has not been set up, the commander may establish one while making provisions for including civil authorities to ensure a unity of effort.

B-71. Not all CMT members are located at the EOC. Some key people may use liaisons to represent them at the EOC. Key personnel can then research and discuss ideas freely, away from the confusion associated with crisis management. The liaison can transmit guidance and answers to the EOC. Some agencies may not be needed in the EOC at first, but they must be prepared for inclusion. The decision to include an agency in the EOC is based on the likelihood that they may have to take an action or a support role on the importance of the agency to the mission.

Appendix C

Urban Terrain Analysis

Of all the environments in which the Army may need to conduct operations, the urban environment confronts commanders with a combination of difficulties rarely found in other environments. The distinct characteristics of the environment result from the combination of man-made features and supporting infrastructure superimposed on the existing natural terrain and the density of civilians in close proximity to combat forces. The human dimension it represents is potentially the most important and perplexing for commanders and their staffs to understand and evaluate. Commanders at all levels must make extraordinary efforts to assess and understand their particular urban environment to plan, prepare for, and execute effective civil disturbance operations.

SIGNIFICANT FACTORS

C-1. Although urban areas throughout the world possess similar characteristics, no two are alike. The sprawl of Los Angeles, for example, bears little physical resemblance to New Delhi. The most significant factors affecting the uniqueness and complexity of each area are its societal characteristics. While the terrain can be extremely complex, information about this aspect, its potential effects on operations, and how it changes over time may be determined with some degree of certainty. However, the human dimension is more difficult to understand and assess, particularly its effect on military operations. The urban environment is not neutral—it helps or hinders. The commander who can best understand and exploit the effects of the urban environment has the best chance of success.

C-2. Whether a metropolitan area or a village, each urban environment has an identifiable system of urban components that constantly change and interact. This system of components consists of the terrain, the society, and the infrastructure that provides the critical link between the two (see *Figure C-1,* page C-2).

C-3. These systems are not separate and distinct categories but instead are overlapping and interdependent. A thorough analysis of these systems contributes to the commander's situational understanding and allows him to develop COAs that apply appropriate resources to control a crowd and prevent a major civil disturbance. This understanding allows commanders to engage decisive points critical to maintaining peace or restoring normalcy to the urban environment.

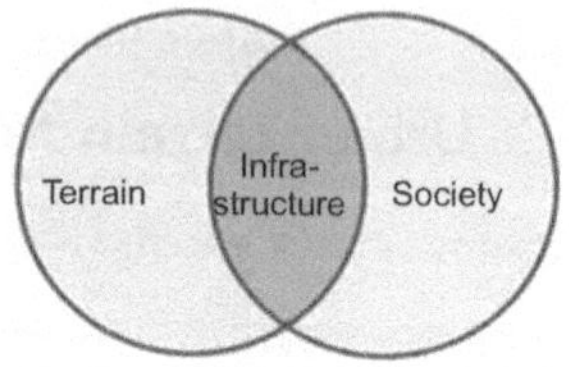

Figure C-1. Keys to Understanding the Urban Environment

C-4. The social aspects and the integrating infrastructure are the most difficult for commanders to understand and assess. These issues require greater dependence on LEAs and local governmental agencies for their information, knowledge, and expertise.

URBAN TERRAIN TYPES

C-5. The physical terrain is the most basic and easily recognizable aspect of the urban area; however, truly understanding it requires comprehending its multidimensional nature, urban and street patterns, urban form and function, and size. The infinite ways in which these factors (see *paragraph C-15)* can intertwine make it impossible to describe a typical urban area. However, these factors provide a framework for understanding the complex terrain within an urban environment. Although the terrain consists of both natural and man-made features, the latter dominates the physical characteristics of the urban environment and has significant effects on military systems and soldiers, tactics, and operations.

URBAN PATTERNS

C-6. The combinations of urban areas, lines of communications (LOCs), and natural terrain result in the identification of four urban patterns (see *Figure C-2*) that will influence civil disturbance operations. An integral part of each urban pattern is the hub or the central urban area around which outlying urban areas radiate. In SOSO, the hub is important, as it is often the economic, political, or cultural center of the surrounding area.

Satellite Pattern

C-7. This pattern consists of a central hub surrounded by smaller, dependent urban areas. LOCs tend to converge on the hub. The natural terrain throughout this pattern is relatively homogenous. Outlying or satellite urban areas often support the principal urban area at the hub.

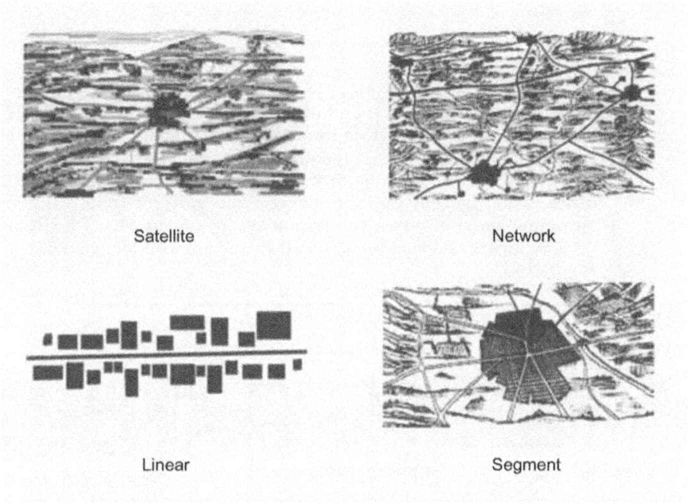

Figure C-2. Urban Patterns

Network Pattern

C-8. This pattern represents the interlocking of the primary hubs of subordinate satellite patterns. Its elements are more self-sufficient and less supportive of each other, although a dominant hub may exist. Major LOCs within a network are more extensive than in a satellite pattern and take a rectangular rather than a convergent form. The natural terrain within a network may vary more than in a single satellite array.

Linear Pattern

C-9. This pattern is a potential subelement of the previous two patterns. It may form one ray of the satellite pattern or be found along connecting links between the hubs of a network. Most frequently, however, this pattern results from the string of minor urban areas aligning a confined natural terrain corridor such as an elongated valley, a body of water, or a man-made communications route.

Segment Pattern

C-10. When dominant natural terrain (such as a river or man-made features [canals, major highways, or railways] divides an urban area, it creates a segmented pattern). This pattern makes it easier to assign areas of responsibility to units. However, if operations are in support of the city LEA, then units should be assigned to a police precinct to avoid crossing several precinct lines, as would happen when assigning units by dominant terrain features.

STREET PATTERNS

C-11. Patterns within the urban area result from the arrangement of the streets, roads, highways, and other thoroughfares. They are a result of the influences of natural terrain, the original designer's personal prejudices, and the changing needs of the inhabitants. The commander must understand the street patterns in the area in which his control force will operate. This will offer force protection, preventing the control formation from being outmaneuvered, flanked, or surrounded by a hostile crowd. Urban areas can display any of the three basic patterns or any combination of the three (see *Figure C-3).*

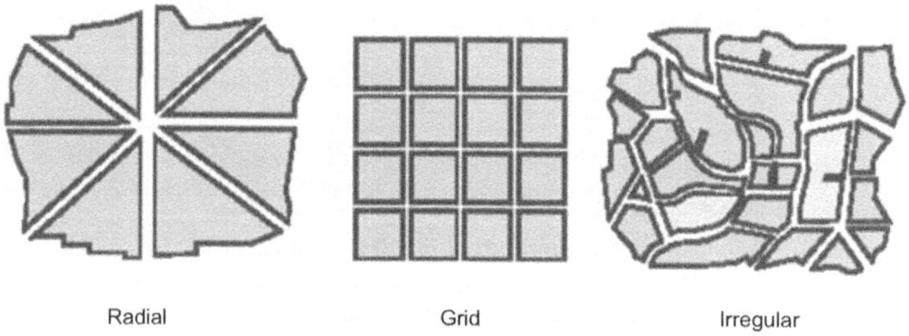

Figure C-3. Basic Internal Street Patterns

Radial

C-12. Societies of highly concentrated religious or secular power often construct urban areas with a radial design with all primary thoroughfares radiating out from the center of power. Terrain permitting, these streets may extend outward in a complete circle or may form a semicircle or arc when a focal point abuts a natural barrier such as a coastline or mountain. To increase mobility and traffic flow, societies often add concentric loops or rings to larger radial patterns.

Grid

C-13. The most adaptable and universal form for urban areas is the grid pattern (lines of streets at right angles to one another forming blocks similar to the pattern of a chessboard). A grid pattern can fill in and eventually take over an original radial pattern. Most urban areas, regardless of the original intent, plan, or vision, emerge from successive plans overlaid on one another.

Areas are well planned to fit with previous plans and others a haphazard response to explosive urban growth. The results may mix different patterns, blend patterns in symmetric combinations, or have no discernable geometric pattern (the irregular pattern).

Irregular

C-14. Irregular patterns can be specifically designed for aesthetic reasons (as in many suburban housing developments). They are sometimes used to conform to marked terrain relief (see *FMs 3-06* and *3-06.11*).

URBAN FORM AND FUNCTION

C-15. Throughout the world, urban areas have similar form and function. In form, urban areas contain like characteristics, readily divisible into distinct sections or areas. Functionally, they tend to be the center of population, finance, politics, transportation, industry, and culture.

- **Core.** The core is the heart of the urban area, the downtown, or central business district. Relatively small and compact, it contains a large percentage of the urban area's shops, offices, and public institutions.
- **Core Periphery.** As the name implies, the core periphery forms the outer edges of the core of the urban area. Generally, it has undergone less change than the core, resulting in buildings of uniform height (generally two to three stories in towns and five to ten stories in larger urban areas).
- **Industrial Area.** Industrial areas generally develop on the outskirts of the urban areas, where commercial transportation is easiest (along airfields and major sea, river, rail, and highway routes). The open/dispersed pattern of the buildings provides sufficient space for large cargoes, trucks, and materials-handling equipment.
- **High-Rise Area.** High-rise areas consist of multistoried apartments, commercial offices, and businesses separated by large open areas such as parking lots, parks, and individual one-story buildings. The automobile, mass transit systems, and improved road networks encourage these areas to grow and function further from the urban core.
- **Residential Area.** Residential areas can be found dispersed throughout the urban area; however, large suburban areas (or sprawl) normally form on the outskirts. The combined population of surrounding suburban areas often outnumbers that of the urban area. Proper and specific suburbs tend toward homogeneity based on ethnicity, religion, economics, or some other social aspect.
- **Commercial Ribbon Area.** Commercial ribbon areas are rows of stores, shops, and restaurants built along both sides of major streets that run through urban areas. These same types of areas often develop along the roads that connect one urban area to another (strip areas).

URBAN SOCIETY

C-16. Although complex, understanding the urban terrain is relatively straightforward in comparison to comprehending the multifaceted nature of urban society. As urban areas increase in size, they generally lose homogeneity; therefore, commanders must understand and account for the characteristics of a diverse population whose beliefs may vary. The behavior of civilian populations within an urban area is dynamic and poses a special challenge to commanders.

C-17. The center of gravity during an urban operation may be the civilian inhabitants themselves. The side that enjoys the support of the population has many advantages. To gain and retain this support, commanders must first understand the complex nature and character of the urban society. Second, they must understand and accept that every military action (or inaction) may influence the relationship between the urban population and Army forces, and by extension, mission success.

C-18. The identification of groups helps commanders focus on specific segments of the urban society to determine their beliefs and needs and how those beliefs and needs motivate them to future action (or inaction). Groups may be categorized based on race, religion, national origin, tribe, clan, class, party affiliation, education level, union memberships, occupation, age, or any other significant social grouping.

C-19. Commanders should consider political, economic, and historical factors in this analysis. These factors usually affect all groups to some extent and often provide the basis for many of their beliefs, needs (actual or perceived), and subsequent behavior. In most cases, training and discipline, grounded in cultural understanding and sensitivity, will help mitigate many of the potential adverse effects resulting from military and civilian interaction.

URBAN INFRASTRUCTURE

C-20. Urban infrastructures are those systems that support the inhabitants, their economy, and their government. They provide the link between the physical terrain and the urban society. Hundreds of systems may exist. Each system has a critical role in the smooth functioning of the urban area. All systems fit into the following five broad categories: communications and information, transportation and distribution, energy, commerce, and human services.

Appendix D

Practical Application

It is always best to derive lessons learned from practical operations, in the field. These are operations where a measure of success was acquired, along with important lessons learned. The following reflects an actual TF application of the doctrine highlighted in the contents of this manual.

OVERVIEW

D-1. The purpose of this manual is to provide TF units and elements with the latest guidance on providing crowd control and preventing civil disturbances.

D-2. Local civilians often gather in towns or along strategic locations throughout an AO to demonstrate grievances or other causes (for example, protesting the apprehension of a fellow citizen, policies of US forces, and policies of government services). Although civilians have the right to assemble, crowd gatherings are predominantly ethnically based, motivated, and influenced. Local civilians can apply for a permit to legally conduct a rally 72 hours in advance as long as their assembly is peaceful and unobtrusive to government services, freedom of movement, or the rights of others. On occasion, intelligence collection can indicate a possibility for violence. In these instances or during spontaneous crowd formations, the TF must provide forces in support of a safe and secure environment or to assist civilian police in preventing unlawful acts. If unlawful acts are committed, detain violators for adjudication under the legal system.

PROCEDURES

D-3. Take precautions to avoid becoming a catalyst to the civil disturbance. During lawful assemblies, the TF must make every effort to treat HN individuals with respect. This means that the TF must engage in peaceful gatherings using interpersonal communication skills. This ensures that demonstrators understand that TF personnel respect their right to lawfully assemble during peaceful demonstrations. When possible, TF engagement of a peaceful assembly should be unobtrusive.

D-4. Do not display civil disturbance equipment or weapons, overtly prepare defenses against a crowd, or aggressively violate crowd space, unless presented with information, intelligence, or negative indicators of pending violence.

NEGATIVE INDICATORS

D-5. Any response to negative indicators should be sequential, measured, and focused on deescalating rather than escalating the crowd.

NEGATIVE INDICATORS OF A POTENTIAL CIVIL DISTURBANCE

D-6. Predictive intelligence is the best indicator of a potential civil disturbance and may provide insight on internal planning, external influences, environmental conditions, or catalyst events that should influence TF planning and response.

D-7. Other indicators of a potential civil disturbance are people beginning to assemble during a scheduled or spontaneous rally. When people begin to assemble, leaders should pay particular attention to crowd dynamics. As the crowd demeanor or dynamics change from peaceful assembly to unruly behavior or unlawful acts, negative indicators should provide an advance warning and prompt a measured response. Negative indicators of a potential change in crowd dynamics and unlawful activities should prompt immediate countermeasures (see *Figure D-1*).

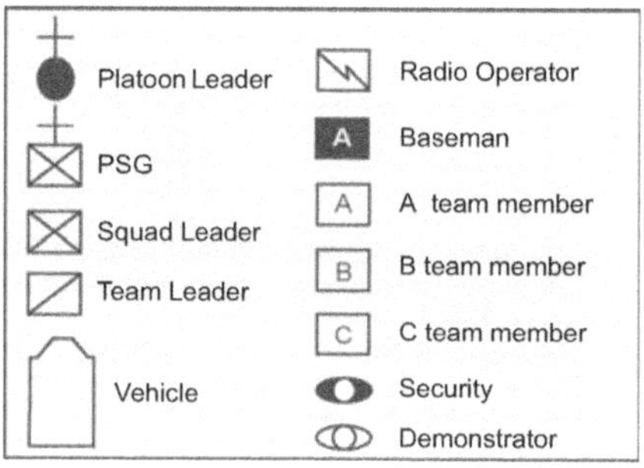

Figure D-1. Symbols for a Formation Element

Orientation

D-8. Crowd dynamics change from people milling around and talking among themselves in isolated groups to a more collective focus concentrated on a single objective (for example, TF personnel, specific agitators or magnets, a government figure, or other activity).

Massing

D-9. The crowd begins to mass or tighten into a large contiguous body from a loose formation to a static congregation located in a more concentrated area. Unchecked, massing can provide anonymity and a collective feeling of invulnerability. This change in group dynamics could provide ideal conditions for a civil disturbance (see *Figure D-2*).

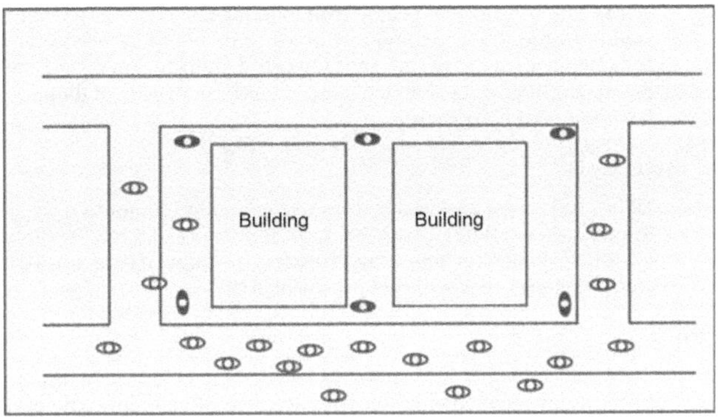

Figure D-2. Crowd Massing

Chanting

D-10. People may begin chanting slogans or yelling ethnically charged rhetoric or obscenities. Because chanting can indicate an increase in crowd intensity, interpreters should be positioned in strategic locations to interpret crowd rhetoric.

Signaling

D-11. Signaling occurs when members of the crowd position themselves to communicate and synchronize crowd actions tied to a plan with specific objectives. Signaling may indicate the intent to escalate the message through the use of violent or unlawful acts.

Blocking

D-12. The crowd may attempt to block public thoroughfares (for example, entrances and exits to public buildings, MSRs, community streets, or other traffic areas that may degrade freedom of movement or similar actions). This escalates the level of attention given to the crowd's message or prevents freedom of movement to opposing persons or groups.

Arming

D-13. Any signs that the crowd is arming itself with objects (for example, bottles, rocks, and sticks) is a clear indication of impending danger to the TF and other civilians. In fact, during a recent civil disturbance, some members of the crowd were seen handing out bottles to others in the crowd. Arming demonstrates intent toward violence is unlawful.

Outside Influence

D-14. The presence of people from outside the community may indicate more sophisticated planning and resources than would otherwise be expected from local community members. Outside people may feel insulated from responsibility for local actions and, therefore, provide a dangerous catalyst for a violent civil disturbance.

Agitators and Instigators

D-15. Agitators and instigators are common elements associated with civil disturbances. They often provide a catalyst to incite the crowd toward violence. Whether spewing rhetoric or committing unlawful acts, their continued presence will fuel potential strife.

Absence of Children

D-16. Recent civil disturbances have shown that even when children were initially present at a rally, they were absent before the civil disturbance erupted with violence. While the presence of children is not a guarantee of a peaceful assembly, the removal of children from a rally may provide early warning for a pending escalation of hostilities or a full-blown riot.

PREVENTION OF CROWD ESCALATION

D-17. If TF personnel believe, based on intelligence or negative indicators, that a crowd is planning to, or may spontaneously, riot they should immediately engage crowd participants using the preventive TTP measures.

D-18. The force executing these preventive TTP can include a number of civil disturbance teams and elements capable of executing the tasks shown in *Figures D-2,* page D-3; *D-3; D-4,* page D-6; and *D-5,* page D-8. These figures demonstrate a full range of crowd control teams and elements, to include camera teams with interpreters, checkpoints, QRF response, riot control formation, NLW teams, snatch and detain teams, blocking teams, and detainee evacuation, as required. *Figures D-2, D-3, D-4,* and *D-5,* illustrate the evolution of a crowd formation and the recommended positioning of teams and elements to engage people as they congregate. Included with each security team is a designated camera or video recorder.

D-19. Depending on the situation and crowd dynamics, it may not be necessary to establish all crowd control teams and elements, but they should be included in planning. Regardless of task organization, teams and elements must be prepared to execute the crowd control and civil disturbance TTP. The order to execute the following TTP should be tailored based on local conditions.

Early Contact

D-20. Engaging people as a crowd forms is essential to prevention. Whether based on intelligence or negative indicators, people should be engaged early, before the crowd masses as shown in *Figure D-2*, rather than waiting until after the crowd masses as shown in *Figure D-3*. Early contact will assist TFs in personalizing their interaction and ensuring that people understand their presence is known, documented, and can be linked to their actions.

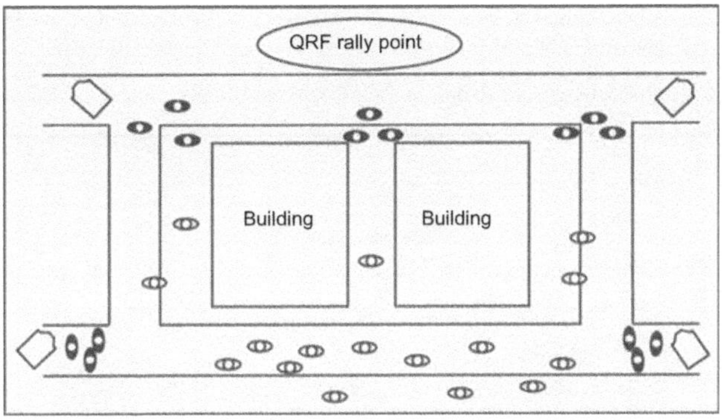

Figure D-3. Early Contact and Unobtrusive Crowd Control Measures

D-21. Cameras are one of the best tools for preventing civil disturbances. The use of cameras to document crowd formations, particularly lens-size groups (small enough to fit in a single photo), will personalize contact and eliminate feelings of anonymity among the crowd. People whose identity has been documented will think twice before committing unlawful acts. Even without documenting names, photo analysis can identify people associated with the civil disturbance, especially those who unlawfully participated as crowd instigators, agitators, or magnets or may commit unlawful acts.

D-22. As people begin to gather in and along streets, camera teams should be deployed as unobtrusively as possible to engage people before they congregate in mass. As depicted in *Figure D-3*, camera teams should consist of two soldiers with an interpreter and be employed in a widely dispersed pattern. Such small, dispersed camera teams are less likely to intrude on lawful rallies or create invasive friction that might become a catalyst to a civil disturbance.

D-23. As the crowd begins to demonstrate negative indicators for a possible civil disturbance (*Figure D-4*, page D-6), camera teams should become more invasive by actively engaging groups to take their pictures. Camera teams should focus cameras on instigators, agitators, or magnets (personnel that garner respect or compliance from the crowd). If a person is purposely avoiding the camera, even if his activities are lawful, the team should take special measures to obtain his picture. Each time a camera team takes a picture of a person or group of people, they must immediately engage them using the warning outlined in *paragraph D-26*.

D-24. Once the crowd has massed and displays negative indicators or begins to riot, camera teams should immediately move behind the riot formation line (see *Figure D-4*). It is essential, however, that camera teams continue to document the event, focusing on people who are committing illegal acts such as arming themselves, throwing objects, and assaulting TF personnel.

D-25. Immediately following a crowd control or civil disturbance operation, photos must be consolidated, processed, and submitted to the TF analysis and control element (ACE) for analysis. A near real-time analysis will provide immediate intelligence for ongoing operations to prevent subsequent civil disturbances or to detain violators from the previous operation.

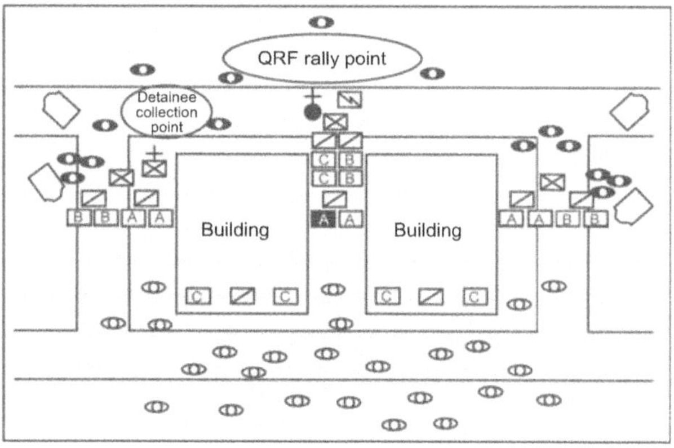

Figure D-4. More Obtrusive Engagement as the Crowd Demonstrates Negative Indicators

Crowd Warning

D-26. Once the team has taken a picture of a person or group of people, they should warn the crowd that as individuals they would be held accountable for their actions. Using an interpreter, the camera team leader will make solid eye contact with the person or group and state the following warning: *"We have your picture. You will be held accountable for your actions. If you commit an unlawful act, you will be detained, if not today, tomorrow. Blocking roads or access to public buildings is unlawful."*

Limit Crowd Size

D-27. Persons seeking to conduct a rally must apply for a permit through civilian authorities at least 72 hours prior to the meeting. If a rally has not been lawfully scheduled through civilian authorities, or if it has, but is demonstrating negative indicators and is turning into a civil disturbance, TF units and elements should attempt to limit the crowd or separate gatherings under 10 people (see *Figure D-3,* page D-5). Walking patrols can assist camera teams in breaking up large groups. Walking patrols should operate in conjunction with camera teams. This will ensure that camera teams can accurately document and provide the aforementioned warning to prevent further escalation of crowd size or behavior.

Clear Routes

D-28. Routes should be considered key terrain and must be controlled by the TF to retain the initiative. Note how a QRF divides in order to clear side streets between protected buildings in *Figure D-4.* Blocking routes or public thoroughfares is not only illegal, but severely restrictive to civil disturbance operations. Routes must remain open to allow movement of TFs and elements, whether it is for camera teams during early contact or for the QRF to respond during an actual civil disturbance. Because blocking routes is illegal, camera teams must document responsible persons for immediate or subsequent detention and legal processing.

Isolate the Area

D-29. As depicted in *Figure D-3,* page D-5, checkpoints should be established to isolate the area. While checkpoints can prevent people from entering the isolated area, they should never prevent people from exiting the area. When trapped, people may escalate the level of violence out of fear, anger, or desperation. When placing checkpoints, leaders should find a location that is both close enough to limit the number of persons being isolated, but far enough away that they will not draw a crowd from people within the isolated area. Checkpoints are essential to preventing others from joining the already gathered crowds. These checkpoints must provide storage room for processing vehicles in both directions; but when limited by space or other considerations, it should simply block traffic trying to enter the isolated area. Improve force protection measures in and around checkpoints as necessary. Strategically placed checkpoints can provide numerous countermeasures to prevent and control a civil disturbance or to identify and detain persons who may have committed unlawful acts during the riot. Use checkpoints to prevent potential participants from joining the crowd or civil disturbance. Checkpoints should also prevent any activity that may commingle ethnic groups. The checkpoint must process all people departing the area by documenting personal and vehicle information with the pictures of all occupants in front of or adjacent to the vehicle.

Establish Lethal Overwatch

D-30. The mentality of "taking the high ground" is just as important in civil disturbance operations as in any other type of operation. Although QRF assets on the ground have both NL and lethal capabilities, it is critical to ensure their safety by deploying marksman/observer teams as shown with the C-team designates on top of the buildings in *Figures D-4* and *D-5,* page D-8. Detain persons committing unlawful acts. The detention and legal processing of persons who commit unlawful acts, against TFs or local laws sends a clear message that all people will be held accountable for their actions. Immediately detain persons who are committing unlawful acts and are isolated or easily accessible, unless the detain operation would create a catalyst event or endanger the apprehension team. If it is not feasible to immediately detain them, a picture with relevant information should be forwarded for analysis and inclusion on the detain list for their detention at a later date and time. Remember, as stated in the warning, they will be detained, if not today, tomorrow. Detainees should be quickly relocated to the forward processing site for detainee evacuation (see *Figure D-5*). Detainee evacuation sites should be behind the control formations to prevent any interference from the local population. Sites should be positioned at two separate locations, such as at the end of a town or located along an MSR. All people who are detained based on

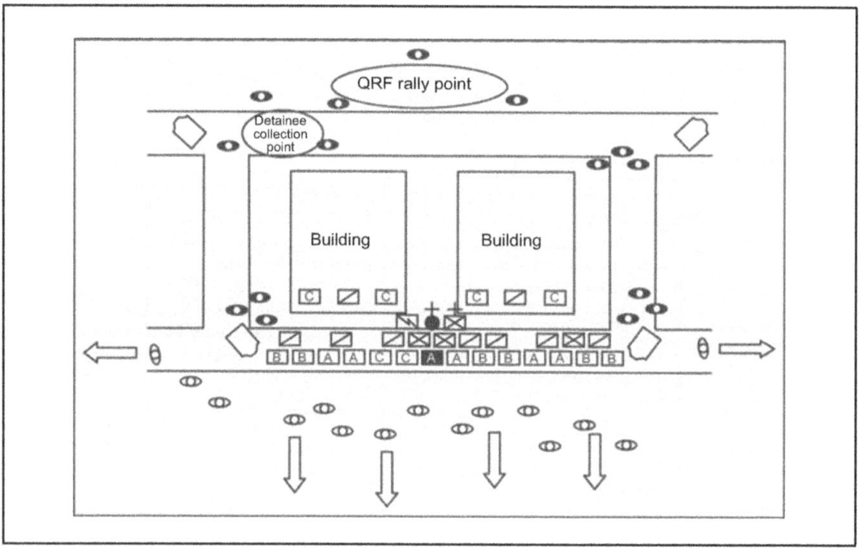

Figure D-5. Full Engagement as Crowd Unlawfully Demonstrates

solid evidence of a crime (for example, pictures of them armed with dangerous objects, blocking a road, assaulting TF members) will be turned over to local authorities for legal processing.

Detain List

D-31. When executing operations based on the detain list, do not show people the list or separate pictures from the list.

Quick Reaction Force Position

D-32. The QRF should be positioned with easy access to decisive points to interdict formations or gathering of crowds, support checkpoints, or provide sustained operations in support of early contact teams. Notice in *Figure D-1*, page D-2, how the QRF is positioned out of direct sight of the crowd until employment becomes necessary. QRF personnel are readily organized, equipped, and easily moved into position to move between buildings and through obstacles (see *Figure D-4,* page D-6).

D-33. At this point, if the crowd ignores verbal orders, NL munitions from appropriate standoff distances could be used to influence and motivate the crowd to comply. Use of NL munitions could prevent the necessity for QRF personnel to come into direct physical contact with the crowd. Early positioning of the QRF could become a catalyst event that may instigate a civil disturbance.

D-34. Assembling in mass (see *Figure D-5*), directly in front of the protected buildings, illustrates to the crowd that there is strength in numbers and the QRF is determined to disperse the crowd from the area. As a reminder, remove the flanking checkpoints that may block the dispersal of the crowd.

Glossary

A2C2	Army airspace command and control
AA	assembly area
AFL-CIO	American Federation of Labor-Congress of Industrial Organizations
AG	Adjutant General
AO	area of operations
ARTEP	Army training and evaluation program
AS	area security
ASV	armored security vehicle
ATTN	attention
BDU	battle dress uniform
CA	civil affairs
CD	combat development
CDC	crowd dispersal cartridge
CFR	Code of Federal Regulations
CINC	commander-in-chief
CMT	crisis management team
COA	course of action
COB	close of business
CONUS	continental United States
COTS	commercial off-the-shelf
CP	candle power
CPU	central processing unit
DA	Department of the Army
DCO	defense coordinating officer
DD	Department of Defense
DM	designated marksmen
DOD	Department of Defense
DOJ	Department of Justice
DOMS	director of military support
EMD	emergency medical dispatching
EMS	emergency medical services
EOC	emergency operations center
FCMT	forced cell move team
FCO	federal coordinating officer
FEMA	Federal Emergency Management Agency
FM	field manual

FRP	federal response plan
FSE	fire support element
FSO	fire support officer
G2	Assistant Chief of Staff, G3 (Intelligence)
G3	Assistant Chief of Staff, G3 (Operations and Plans)
G5	Assistant Chief of Staff, G5 (Civil Affairs)
G6	Assistant Chief of Staff, G6 (Information Management)
GOTS	government off-the-shelf
GRM	graduated response matrix
HIV	human immunodeficiency virus
HMMWV	high-mobility multipurpose wheeled vehicle
HN	host nation
HQ	headquarters
I/R	internment and resettlement
INIWIC	Interservice Nonlethal Individual Weapons Instructor Course
IO	information operations
IPB	intelligence preparation of the battlefield
JMC	joint military commission
JTF	joint task force
L&O	law and order
LBE	load-bearing equipment
LBV	load-bearing vest
LEA	law enforcement agency
LOC	lines of communication
LOS	line of sight
LP	listening post
MACDIS	military assistance for civil disturbance
MANSCEN	Maneuver Support Center
MCCM-GE	modified crowd control munition-ground emplacement
MCM	Manual for Courts-Martial
MDMP	military decision-making process
METL	mission essential task list
METT-TC	mission, enemy, terrain, troops, time available, and civilian considerations
MI	military intelligence
mm	millimeter
MO	Missouri
MOOTW	military operations other than war
MOUT	military operations on urbanized terrain
MPI	military police investigator
MRE	meal, ready to eat

MS	mobility support
MSCA	military support to civil authorities
MTP	mission training plan
MWD	military working dog
NCO	noncommissioned officer
NEO	noncombatant evacuation operations
NG	National Guard
NL	nonlethal
NLCS	nonlethal capabilities set
NLW	nonlethal weapons
OC	Oleoresin Capsicum
OCONUS	outside continental United States
OES	office of emergency services
OIC	officer in charge
OP	observation post
OPFOR	opposing forces
OPORD	operations order
PA	public address
PAO	Public Affairs Office
PIO	police intelligence operations
PM	provost marshal
PLT	platoon
PMO	Provost Marshal Office
POMSO	plans, operations, and military support officer
PPCT	pressure point control tactic
PPE	personal protective equipment
PSG	platoon sergeant
PSYOP	psychological operations
PT	physical training
PVAB	portable vehicle arrest barrier
QRF	quick reaction force
RCA	riot control agent
ROE	rules of engagement
RUF	rule for the use of force
S2	Intelligence Officer (US Army)
SCO	state coordinating officer
SCRAG	senior representative of the attorney general
SITREP	situation report
SJA	Staff Judge Advocate
SOFA	status of forces agreement

SOMA	status of mission agreement
SOP	standing operating procedure
SOSO	stability operations and support operations
SQD	squad
STARC	state area command
STP	soldier training publication
STX	soldier training exercise
TA	Training Area
TF	task force
TLP	troop leading procedure
TM	team
TPT	tactical psychological operations team
TRADOC	United States Army Training and Doctrine Command
TSOP	tactical standing operating procedure
TTP	tactics, techniques, and procedures
UCMJ	Uniform Code of Military Justice
UN	United Nations
UNGA	United Nations General Assembly
UNHCR	United Nations High Commissioner for Refugees
UNSC	United Nations Security Council
US	United States
USAMPS	United States Army Military Police School
VCR	video cassette recorder
VIP	very important person
WMD	weapons of mass destruction
WTO	World Trade Organization

Bibliography

AR 380-10. *Foreign Disclosure and Contacts With Foreign Representatives.* 6 June 2003.

AR 380-13. *Acquisition and Storage of Information Concerning Nonaffiliated Persons and Organizations.* 30 September 1974.

Code of Federal Regulations.

Constitution of the United States.

DA Form 2028. *Recommended Changes to Publications and Blank Forms.*

DA Form 4137. *Evidence/Property Document.*

DD Form 2708. *Receipt For Inmate or Detained Person.*

DOD Civil Disturbance Plan 55-2. *Operation Garden Plot.* 15 February 1991.

DOD Directive 3025.12. *Military Assistance for Civil Disturbances.* 4 February 1994.

DOD Directive 3025.15. *Military Assistance to Civil Authorities.* 18 February 1997.

DOD Directive 5525.5. *DOD Cooperation With Civilian Law Enforcement Officials.* 15 January 1986.

Executive Order 12333. *United States Intelligence Activities.* 4 December 1981.

Fiscal Year 1989 Act.

FM 3-06. *Urban Operations.* 1 June 2003.

FM 3-06.11. *Combined Arms Operations in Urban Terrain.* 28 February 2002.

FM 19.10. *Military Police Law and Order Operations.* 30 September 1987.

FM 3-19.4. *Military Police Leaders' Handbook.* 4 March 2002.

FM 3-19.40. *Military Policy Internment/Resettlement Operations.* 1 August 2001.

FM 3-21.5. *Drills and Ceremonies.* 7 July 2003.

FM 3-22.40. *Multiservice Tactics, Techniques, and Procedures (MTTP) For Tactical Employment of Nonlethal Weapons (NLW).* 15 January 2003.

FM 7-0. *Training the Force.* 22 October 2002.

FM 7-1. *Battle Focused Training.* 15 September 2003.

FM 27-100. *Legal Support to Operations.* 1 March 2000.

Judiciary Act of 1789.

MCM.

National Defense Authorization Acts.

NGR 500-1/ANGI 10-8101. *Military Support to Civil Authorities.* 1 February 1996.

Posse Comitatus Act. 1878.

STP 19-95B1-SM. *Soldier's Manual for MOS 95B Military Police Skill Level 1.* 14 January 2003.

STP 19-95B24-SM-TG. *Soldier's Manual and Trainer's Guide for MOS 95B, Military Police, Skill Levels 2/3/4.* 2 December 2002.

STP 19-95C1-SM. *MOS 95C, Corrections Specialist, Skill Level 1 Soldier's Manual.* 30 October 2003.

STP 19-95C24-SM-TG. *MOS 95C, Corrections Specialist, Skill Levels 2/3/4, Soldier's Manual and Trainer's Guide.* 30 October 2003.

Uniform Code of Military Justice.

USC, Title 32. *National Guard.*

USC, Title 10, Chapter 18. *Military Support for Civilian Law Enforcement Agencies.*

USC, Title 10, Subtitle A, Part I, Chapter 15, Section 331. *Federal Aid for State Governments.*

USC, Title 10, Subtitle A, Part I, Chapter 15, Section 332. *Use of Militia and Armed Forces to Enforce Federal Authority.*

USC, Title 10, Subtitle A, Part I, Chapter 15, Section 333. *Interference With State and Federal Law.*

USC, Title 10, Subtitle A, Part I, Chapter 15, Section 334. *Proclamation to Disperse.*

USC, Title 18, Part I, Chapter 67, Section 1385. *Use of Army and Air Force as a Posse Comitatus.*

Index

Made in United States
Cleveland, OH
05 November 2026

47673670R00191